MISSION
MOON
3-D

David J. Eicher
Brian May (Creative Director)

"Let us hope that our grandchildren at our age can look back and say, 'The 20th century was a century of advancement and improvement in technology, and the 21st century was a century of advancement and improvement in human character.'"

— *Neil Armstrong*

"While it was conflict between two competing superpowers, two ideologies, that propelled me into space, in the moment when I looked back at Earth it struck me very forcefully that our planet is home to just one human race."

— *Alexei Leonov*

MISSION
MOON
3-D

The MIT Press, Cambridge, Massachusetts

THE
London Stereoscopic Company,
LTD.

Published in 2018 by The London Stereoscopic Company
© The London Stereoscopic Company 2018
Text © David J. Eicher 2018
David J. Eicher has asserted his right to be identified as the author
of this work in accordance with the Copyright, Design and Patents Act 1988 (UK).

The London Stereoscopic Company

Director: Brian May
Publisher and Chief Editor: Robin Rees
Art Director: James Symonds
Archivist: Denis Pellerin
LSC Manager: Sara Bricusse
PR Manager: Nicole Ettinger
Office Manager: Sally Avery-Frost
Website: Phil Murray

Project Specialists

Stereoscopic Supervisor and Cover Design: Brian May
Astro 3-D Compositor and Researcher: Claudia Manzoni
Original Concept: Glenn Smith
Starmus Festival Director and Project Consultant: Garik Israelian
Soviet Space Consultants: Brian Harvey, David Shayler
Consultant: Steve Young
Proofreaders: Sarah Tremlett, Mandy Bailey
MIT Press Cover Design: Molly Seamans

Management: Jim Beach

Published in North America, Australia, and New Zealand by
The MIT Press
Cambridge, MA 02142
Library of Congress Control Number: 2018946828
MIT Press ISBN 978-0-262-03945-1

A catalogue record for this book is available from the British Library.
UK ISBN 978-1-9996674-0-5
Book printed and bound in China by Jade Productions.
OWL Stereoscope designed by Brian May.
First printing 2018.

www.LondonStereo.com
www.BrianMay.com
BrianMayForReal
@DrBrianMay
@LondonStereo

CONTENTS

FOREWORD
BY CHARLIE DUKE

A grand total of a dozen men had the amazing experience of walking on the Moon. All Apollo astronauts, they had the glorious vista of looking back on a small Earth, with all of its peoples, floating in an inky black sky. They walked — and sometimes drove — across a lunar surface that was either strangely dark gray, a powdery expanse, or awash in bright sunlight, making it bright gray. We saw occasional glimpses of color in the Moon rocks as we explored on our stays on another world, all of which lasted three days or less.

Between 1969 and 1972, this club of 12 enjoyed a unique experience. The members are Neil Armstrong, Buzz Aldrin, Pete Conrad, Alan Bean, Alan Shepard, Edgar Mitchell, David Scott, James Irwin, John Young, Gene Cernan, Jack Schmitt, and myself. I was blessed to be the Lunar Module Pilot on Apollo 16, and became the tenth and youngest person ever to walk on the Moon. The club was small, and now is even smaller.

In later years, I've had the fabulous experience of renewing many of those acquaintances through my participation at the Starmus Festival. It was there also that I met Brian May, and was delighted to learn about how his passion for stereoscopy has led to this unique book. Only four of us are around who walked on the Moon, but now — through the images in this volume — you will have the chance to get as close to walking on the Moon as possible, without actually traveling there. The three-dimensional views of the lunar surface that Brian and his team have created are amazing, and give a good approximation of what it's like to really be there.

The Apollo 16 mission in the spring of 1972 was a wild adventure. Along with John Young, I spent three days exploring the lunar surface in the region of the Descartes Highlands, west of Mare Nectaris. We collected many samples and studied ejecta, showing that it came from impacts rather than volcanism. On stepping onto the Moon, my first words were, "Fantastic! Oh, that first foot on the lunar surface is super, Tony!," referring to Tony England, who was in Houston.

John and I conducted three extravehicular activities, or moonwalks, during the three days of our adventure. We were fortunate to span a pretty extensive area due to the fact that we had a lunar rover at our disposal. We explored lots of regions, names for which included Stone Mountain, South Ray Crater, Flag Crater, the Cinco Craters, the Vacant Lot, Palmetto Crater, and House Rock. The memories of those days are still vivid, and amazing to me. How I wish that we had this kind of exploration still going on in our current world — what that would do for science!

You can read much more about Apollo 16, and all the other missions, and indeed the race we were in to get into space as quickly as we could, in these pages. I hope the story of these fascinating times, in which humans took their first steps out into space, along with these amazing and unprecedented images, will entertain you for a long time to come.

April 21, 1972: Charlie Duke, Lunar Module Pilot of Apollo 16, on the surface of the Moon collecting geological samples.

June 20, 2017: Charlie Duke, guest speaker at the Starmus Festival in Trondheim.

PREFACE
BY DAVID J. EICHER

When John F. Kennedy called Americans to land a man on the Moon and return him safely to Earth, before the end of the 1960s, it stunned the American government, surprised the people, and shook the world. For a time such a thing seemed overly ambitious at a minimum, and unrealistic to many. And yet it happened. Kennedy's vision shows us what can be done in constructive ways if we only can come together to work as one. And the more cynical might add that the value of a little international competition is not to be underestimated!

To my great pleasure, I ended up working on this book with a group of friends. It all began with the Starmus Festival, a unique happening in the world of science. Astronomer Garik Israelian created Starmus — stars and music — in 2011. It is a science event that has brought together many of the world's greatest minds — scientists, astronaut-explorers, Nobel Prize laureates, musicians and artists. The Festival was born out of the friendship between Garik and Brian May, founding member and guitarist of Queen, and astrophysicist. It was my good fortune to get to know Garik and Brian around the time of the first Starmus, and by 2014 they added me to the Starmus board of directors, where I could gleefully hang about with many incredible people, including boyhood heroes such as Buzz Aldrin, Alexei Leonov, and Charlie Duke.

The book you are holding really originated as a concrete idea when the fourth Starmus Festival took place in Trondheim, Norway, in 2017. Although Brian was busily away from this Starmus, touring with Queen, a number of us met up during the Festival. Denis Pellerin was engaged in making 3-D photographs of many of the astronauts on site, for the London Stereoscopic Company. Brian's publisher Robin Rees was on site overseeing the publication of the Starmus books. Our good friend Glenn Smith, a world expert on planetaria, simply said, "Hey, someone should do a book for the 50th anniversary of the Apollo 11 Moon landing."

The idea immediately struck us like a bolt of lightning. It occurred to us that such a project would bring together all of our skills and passions. Thanks to Garik and the Starmus Festival, we all had "backstage passes" and the ability to connect with all the right parties, and thanks to Brian we had the vehicle, the London Stereoscopic Company, with which to publish such a tribute book. Robin declared: "You write it, Dave!"

Brian embraced the idea warmly, and was 100 percent for it. He planned the book to align with one of his great passions, stereo photography, and to present as many images as possible in glorious 3-D, as never seen before. He enlisted his collaborator in all things astronomical, Claudia Manzoni, who is a stereo photo whiz. Brian and Claudia immediately began researching and creating stereo photos from the era.

David J. Eicher, with an OWL stereoscope.

The interconnections between members of the group quickly grew stronger. There was a feeling that *Mission Moon 3-D* was meant to be, and that we had the perfect team to pull it off. The book took on a life of its own. It has been an incredible and amazing joy to work on this project with this splendid group of folks. As we experience the 50th anniversaries of missions, to many, the Apollo era seems unthinkably long ago. And yet ties to the time are very real. My wife Lynda was born on the very day JFK delivered his call to action, May 25, 1961. Beyond spaceflight, even, the ties to the past are closer than we might think. My father, John Eicher, lived to be 95 years old, passing in 2016. He grew up in Dayton, Ohio, and lived two doors down from the niece of the Wright Brothers, Ivonette Wright Miller. Every once in a while, in the 1930s, Orville Wright would pull up in his big, black touring car with acetylene lights and have a chat with my dad. Connections to the invention of powered flight itself are not too far away.

Some 150 3-D photos of the space race lie scattered throughout this book, brought to life with Brian's OWL stereoscopic viewer, which is included with this volume. If you are an experienced viewer of stereo photos, you may be able to "free view" the images, relaxing your eyes and merging them into one without the viewer. In any case, you are in for hours of exciting exploration with this work, and also included are many mono images that illustrate the incredible story of this time and its herculean effort.

As I've written this book, I received a great deal of support from my family – wife Lynda and son Chris. They have helped me by allowing me to take on such a project and also providing ample encouragement to keep the thing going until liftoff.

"They said it couldn't be done, so we didn't even try!" This tongue-in-cheek slogan, sometimes applied to Apollo, worked out pretty well in the end. We did try. We hope that you will enjoy this effort, a retelling of the space race on its important anniversary that shows the Moon landing experiences like never before. We hope you will read it and reread it, and remember those glory days of space exploration with great fondness.

Upward!!

INTRODUCTION BY BRIAN MAY

Brian with a copy of *The New York Times* from July 21, 1969.

Fifty years ago, men travelled a quarter of a million miles through space, and set foot for the first time on our planet's Moon. Today, this outrageously bold adventure seems as shiny and new as if it were yesterday. It still captures the imagination of kids of all ages – witness, for instance, the ever popular LEGO model of a Saturn rocket and the capsule that took men on that epic voyage. It seems that the Apollo rockets are set to stand alongside the historic Spitfires and Hurricanes of World War II as the hardware of perennial fantasies for children, as the stories are retold to new generations – legends almost too daring to be true.

And so in this book we re-tell the incredible story of the space race, the astronauts, the cosmonauts, the spacewalkers and the moonwalkers – their tribulations and

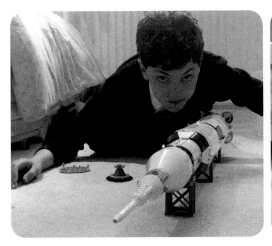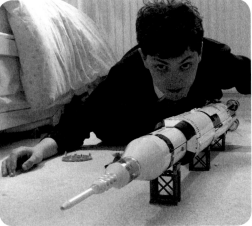

Twenty-first century LEGO model of a Saturn rocket, with a young modeller.

View-Master stereo of the Apollo 11 press conference, before the launch.
From left to right: Neil Armstrong, Buzz Aldrin, and Michael Collins.

triumphs – and the technological marvels that propelled them. But we will take you on this journey in a way that's never been possible before, even though the trail is half a century old, through a narrative which gives us perspectives from *both* sides of that scramble to reach the Moon, including stories which could never be told until now. Also, we will see it through 150 newly-created stereoscopic pairs of photographs, viewable right here in these pages in glorious 3-D, using the OWL stereo viewer supplied with the book.

David J. Eicher is a renowned writer on all things space – the Editor-in-Chief of the world's bestselling astronomical periodical – *Astronomy* magazine. David has written a brand new text for this publication which gathers a mountain of information from the United States and Russia (or the USSR as it was then), and he reveals in detail how the lunar exploration teams of both nations pushed the boundaries of their technology to the limits; how they developed new propulsion and guidance techniques at a rate which had been considered impossible only a few years before. But David also weaves the human story strongly into his narrative – how these brave men shaped their destiny, conquered their fears and made 20th century history. Uniquely, in this account, David has also interwoven the social story which paralleled the Moon missions. Garik Israelian's recent Starmus conferences have emphasised the very strong connections between art and science in space exploration, and in particular the musical associations found in both performers and 'audience'. So in this book there is one more thread of consciousness – the parallel development of rock music, the music of the space generation.

But what of the images? You will see as you turn these pages some 'mono' photographs which need no special apparatus to view. But interspersed among them are the side-by-side pairs of photographs which will give you a uniquely powerful new three-dimensional perspective – which Charlie Duke himself has described as making him feel almost as if he were back on the surface of the Moon, as he was in 1972.

Instructions on how to get the best immersive experience from these 3-D pictures can be found inside the back cover of the book. Take a few moments to master the viewing technique, and you'll be well rewarded.

But how did this 3-D record come about? How is it possible to see the Moon missions in 3-D at all? Surely the astronauts didn't carry 3-D cameras to the Moon? And why has this experience never been possible before, since evidently all the photographic material from these expeditions must have been available to the public for the last 50 years? Well, the answers are many and various!

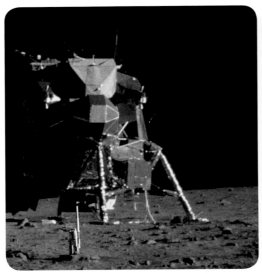

Perched on a flat rock in front of the *Eagle*, Thomas Gold's stereoscopic camera (ALSCC) proudly stands in this stereo picture derived from Neil Armstrong's photographs. Neil was in charge of taking stereoscopic close-up photographs of the lunar soil with the ALSCC, and left it there while he deployed the lunar retroreflector used in laser ranging experiments to accurately determine the distance between the Earth and the Moon.

Firstly, it's not quite true that no stereo camera ever reached the Moon. Apollos 11, 12 and 14 carried a dedicated close-up 35 mm film stereo camera called the ALSCC (Apollo Lunar Surface Close-up Camera). Designed by Prof. T. Gold of Cornell University, the camera was mounted on a walking stick, enabling the astronauts to photograph a three-inch square area of dust and small rocks with flash illumination and at a fixed distance, by pressing the assembly onto the spot and pulling a trigger. We have included some of the results here (see pages 98, 134).

But throughout the whole program, there was never a camera available for taking 3-D pictures of people or landscapes or space vehicles. If Apollo had taken place in Victorian times, the 1850s, there is no doubt that the crew members would have been equipped with stereo cameras – they were one of the technological marvels of the age. But in the 1960s, even though stereoscopy would have enabled them to capture valuable extra data, stereoscopy was not considered a priority.

However, 3-D was not entirely ignored. Astronauts on the Apollo missions were trained in the art of *sequential* stereoscopy. So, for instance, when Stuart Roosa in Apollo 14 was circling the Moon while his colleagues were walking on the Moon, he deliberately took pairs of pictures of the same lunar surface features, separated by a few seconds, so that the pictures could be later assembled into stereo views. You will see some of these, newly processed and 'polished', from Apollo 8 onwards.

But perhaps at this point we need to explain what 3-D actually is, and how it works. Evolution has furnished humans with two eyes, not one. Evidently there was a very good reason for this! We evolved with twin optical sensors about 2½ inches apart, supplying our brains with two slightly different views of our surroundings every second of our waking lives. Our brains developed the ability to combine these two views instantaneously to form a three-dimensional image in our minds. It's a miracle (if 'miracle' is an appropriate term to use about evolution!) that we generally never stop to appreciate, but this perceived image, with depth and solidity, gives us a perception of what is around us that is vastly more informative than a mere 'flat' view from one eye. This undoubtedly contributed significantly to our survival, since we were able to instantly assess the size and shape of a dangerous predator or potential food source.

This amazing ability to see in 3-D is called *stereopsis*, and it was first understood in the 1830s by an eminent English scientist – Charles Wheatstone, who invented an optical device which could reproduce that sensation of depth. He called it a *stereoscope*.

So to obtain a 3-D picture of anything, the name of the game is to create two different views of the same object, captured by two cameras – at a certain distance apart known as the *baseline* of the stereo view. For everyday subjects such as people or landscapes, a 'normal' baseline of a few inches is fine. But to capture depth information in astronomical-sized objects, astronomical-sized baselines are needed.

For example, as the early space missions left the vicinity of the Earth for the first time, they were able to see our planet spin on its axis, and, taking photographs of it at intervals, they effectively captured the Earth from a number of different viewpoints. Combining two of these views gives the baseline we need, and the resulting stereo pair makes an impressive 3-D picture of the Earth (see pages 100, 119).

This is a Swedish-made Hasselblad 70 mm lunar surface super wide-angle camera as used in the Mercury, Gemini and Apollo space programs. The cameras were electrically powered and semi-automatic, operated by a trigger in the camera's handle. They were left on the Moon to save weight, and only the film magazines were brought home – including this one from Apollo 12.

The same applies to pictures the astronauts took of rotating space capsules from their windows. In the heat of the moment, Neil Armstrong, for instance, showed amazing coolness in capturing such pictures in lunar orbit, and we have been able to combine them to make stereo views. What we are saying, from this point on, is that the astronauts *sometimes* took stereo views by 'accident' since their minds were focused on other things. We're fortunate that on the surface of the Moon they often took multiple exposures of a subject, using the Hasselblad cameras they were equipped with, and if they were moving around between shots, we have often found that we can make the stereo pairs we need. However it's not quite that simple (of course!). Many stereos have been made in the past from these pictures, but usually they were assembled without applying any corrections for distortions, or lighting and shadow changes, or for the problem that the subjects were not always kind enough to stay still between the left and right shots. Moreover, these stereos were usually presented in the old 'red and green' *anaglyph* format. Using the 1950s red and green spectacles on these lunar anaglyphs gives some 3-D effect, but most of the colour information is lost, and although the format is forgiving, the murkiness of the view, coupled with the 'retinal rivalry' caused by the errors, very quickly induces fatigue in the person viewing the pictures.

There is not room in this section to detail all the techniques we've used to correct and polish the 'accidental' stereos we've found in the NASA archives – but the new digital medium has made many things possible that were previously next to impossible. *Mission Moon 3-D* team member Claudia Manzoni has spent uncountable hours trawling through the digitised archives for stereo opportunities which had preciously been missed, and my own obsession with processing in Photoshop in most cases did the rest. Denis Pellerin, my London Stereoscopic Company chief researcher, also contributed many of the images, and his life-long interest in the Moon missions has enabled him to do much of the coordination in the assembly of the book. James Symonds, our venerable designer, veteran of six previous LSC 3-D books, has magnificently shaped the entire book, including meticulously preserving the 'stereo windows' we created. Yes, stereo windows are another story and we recommend anyone who is drawn to go 'deeper' into stereoscopy to visit our LondonStereo.com website and/or peruse our previous publications.

Returning to the creation of the stereo images ... we have also drawn on some other sources. Back in the 1950s and 1960s there *was* an outlet for 3-D pictures ... in the highly successful View-Master viewers many of us grew up with, and View-Master published at least two sets of reels devoted to the US space missions. We have gratefully reproduced a handful in this book, and we thoroughly recommend picking up these original treasures if you come across them.

This classic Bakelite View-Master stereo viewer was at the time the ultimate way to enjoy the exploration of space in 3-D. Reels on different themes could be viewed by dropping them into a slot at the top of the viewer. Clicking a lever changed the subject.

We have occasionally used movie films of the day to provide stereo opportunities... sequential stereos often become possible to extract, especially when the camera tracks sideways. And if all else fails, there is one last resort! It *is* possible to do 'conversions' from mono pictures to stereo views. This involves a great deal of skill and the addition of a lot of assumed information about the depths of the original scene. 'Conversion' is used these days to create most of the 3-D movies we see in the cinema. It's done frame by frame using highly sophisticated algorithms and depth maps. We are grateful to *William3D* for just a few of the stereos you will encounter here ... notably a fine adaptation of our picture of Charlie Duke on the Moon.

Finally we must acknowledge the other two members of our team, fulldome planetarium wizard Glenn Smith, and science publishing guru Robin Rees. It was in a conversation between Glenn and Robin that Glenn first asked if we, the London

Alexei Lenov, Brian May, Neil Armstrong, at Starmus 2011.

Stereoscopic Company, had considered creating the first stereoscopic 3-D book dedicated to the Moon missions. From the initial discussion this inspired, *Mission Moon 3-D* has proceeded at an amazing speed, considering the complexity it demanded. It's thanks to the dedication of the entire compact team that this book has turned out to be a creation to be proud of.

Yes, we have used 'every device' to thrill you in this publication. But most of what you will experience in your OWL stereoscope is entirely *real*. We hope you will enjoy this 'virtual' trip to our nearest neighbour in space, and feel that you can continue to visit, and experience its wonders.

Brian tries his hand with a 1960s' lunar inspired game, Johnny Astro.

On April 12, 1961 Yuri Gagarin became the first human being to travel in space. Orbiting the Earth for 108 minutes in his Vostok 1 spacecraft, this astonishing achievement caught the world by surprise and jolted the Americans into action – the space race was on!

1 THE RACE BEGINS
(SPUTNIK 1, VOSTOK 1)

It began as a rare occurrence. In the United States Capitol, braced at the rostrum of the US House of Representatives, America's youthful President John F. Kennedy (also known as Jack) delivered a declaration. The day was May 25, 1961, and Kennedy addressed a joint session of Congress. Normally such an address came once a year, the "State of the Union" delivered in early January, but in Kennedy's words these were "extraordinary times" confronting Americans with "extraordinary challenges." So the President assembled both houses of Congress to shine a spotlight onto key ideas at a crucial moment.

President Kennedy speaks before a crowd of 35,000 people at Rice University's football field on September 12, 1962.

Kennedy spoke sternly to Congress, a battery of microphones in front of him, and flanked to the rear by Vice President Lyndon Johnson and Speaker of the House Sam Rayburn. The President painted America as the bastion of freedom and other forces in the world more darkly, and asked for increases in national defense and military aid to other nations. He announced a program to build and equip fallout shelters in the event of nuclear war. Senators and representatives — as well as the broadcast audience — listened intently, if not glumly, some still adjusting to the growing sense of danger and risk they felt in the world.

Late in the speech, Kennedy turned to his longest portion of the document devoted to any single topic — space. Following a prelude, Kennedy uttered the line that resonated all over the world: "I believe that this nation should commit itself to achieving the goal, before this decade is out, of landing a man on the Moon and returning him safely to the Earth. No single space project in this period will be more impressive to mankind, or more important for the long-range exploration of space; and none will be so difficult or expensive to accomplish."

Summarizing the call to action, Kennedy concluded: "It is a most important decision that we make as a nation. But all of you have lived through the last four years and have seen the significance of space and the adventures in space, and no one can predict with certainty what the ultimate meaning will be of mastery of space."

The call to action to the Moon stunned the Congress, the nation, and the world. Even Kennedy's science advisors, while they believed that a successful program of space probes could improve the political standing of the United States, doubted that Americans could beat the Soviets in a race to the Moon. The stakes were high, and they were about to be raised to an even higher level.

Of course the whole saga of the exploration of space played out over the canvas of the "cold war", a period of stiff political tensions between the two post-World War II superpowers, the Soviet Union and the United States. Fast friends during the war, at least in shared military goals, the two nations drifted badly apart following Harry Truman's 1947 doctrine that offered copious aid to nations threatened by Soviet expansionism. And of course the whole balance of power reflected the furiously growing programs of nuclear energy on both sides that expanded alarmingly after the use of the atomic bombs to end the war. So the relationship between the two nations was dramatically fragile and hinged on both nations aiming to expand influence, which fueled the cold war as a self-fulfilling prophecy.

As Kennedy proposed a journey to the Moon, the greatest crisis of the cold war to date was just in the rearview mirror. In April 1961 the US Central Intelligence Agency (CIA) launched an ill-fated invasion into Cuba, where it hoped to overthrow pro-communist dictator Fidel Castro. Some 1,400 paramilitary troops moved on Cuba via Guatemala, supported by eight CIA-supplied bombers. On April 16, the invasion party landed on a beach at the Bay of Pigs and met a counteroffensive by revolutionary militia. The invaders lost the initiative, US involvement in the operation became known, and worldwide sentiment turned against the operation. Kennedy canceled further air cover and the exercise stalled, with the invaders surrendering after three days and ending up in Cuban prisons. The US failure made Castro a hero and cemented his leadership position in Cuba, a communist nation a scant 160 kilometers (100 miles) from the US mainland.

So to Americans, the call to go to the Moon was set against a backdrop of competition, military alarm, and fear. And some of that fear originated with Soviet successes that placed the Russian exploration of space far ahead of the United States.

1961: cosmonaut Yuri Gagarin, the first man in space, being congratulated following his historic flight; his Vostok 1 spacecraft completed an orbit of the Earth in 108 minutes.

Almost at the same time as the Bay of Pigs fiasco, the Soviet Union scored the first major victory in the quest for space. The heroic cosmonaut Yuri Gagarin, aged 27, completed an orbit around Earth on April 12, 1961, becoming the first human being in world history to travel into space. The amazing achievement stunned and excited the world, and billions celebrated. Gagarin orbited Earth in a Vostok spacecraft for 108 minutes, and the accomplishment seemed to come out of nowhere, having been planned and held in secret beforehand. The gateway to space was now open. The Soviets were leading the way. And the United States, cast in the psychology of the cold war, was, well, left out in the cold.

And so Kennedy's speech, with its call to arms to go to the Moon, was something of a shock reaction to the Soviet success, and the expectation and belief that the Russian program was far ahead of the American program, which of course it was. The Americans could do nothing, it seemed, but attempt to play catch-up.

The whole affair had started in 1957, when suddenly the Soviet Union successfully launched the first Earth-orbiting satellite, Sputnik 1. This milestone sent political shock waves around the world. Could the Russians now drop bombs directly down on anything and everything from space? Political panic outstripped rationality, but that was the nature of the cold war.

In reaction, the United States established the National Aeronautics and Space Administration (NASA) in 1958. Lagging the Soviets once again in what became a fevered race, the Americans launched astronaut Alan Shepard into a suborbital flight on May 5, 1961, as the first action in its Mercury program. Again the sense was that the Americans accomplished too little, too late. The political struggle to show which nation could be the leader in space exploration instigated what was to become a race to the Moon. Politics morphed into a dysfunctional partner with science in this effort — a necessary evil. Without the political drive for psychological dominance — "my rocket is bigger than your rocket!" — no one would have tried to go to the Moon. It's sad, but true. The scale of the engineering project Kennedy called for would be exceeded in scope and expense only by the Manhattan Project and the construction of the Panama Canal.

The young American President who felt backed against the wall, who called for a journey to the Moon, was unique in American history. Kennedy delivered his

On his return from space, Gagarin instantly became a huge international celebrity. He was awarded many medals and titles, and is seen here in a motorcade on his way to receive his nation's highest honor, Hero of the Soviet Union, from Nikita Kruschev. Only seven years later, Gagarin was to meet his end prematurely, but not in space.

address to the joint session of Congress just four days shy of his 44th birthday. The American people, and to some degree the world, had fallen for this youthful, optimistic, energetic, and extremely bright leader — the first US President to be born in the 20th century.

Kennedy was the scion of a political family from Massachusetts, the son of Joseph P. Kennedy, businessman, investor, and politician who had served as US Ambassador to the United Kingdom under Franklin Roosevelt. The elder Kennedy groomed his large Catholic family for politics. The leading and oldest son, who was targeted for a major political run, Joseph Kennedy Jr., was killed in action as a bomber pilot in the Second World War. The senior Kennedy's high political aspirations for the family then fell to John F. Kennedy, the next son in line. After commanding a patrol boat in the war, John F. Kennedy took up the political career, being elected to the US House in 1947 and the Senate in 1953. The Democratic Senator from Massachusetts gained fame as he wowed the public and fellow politicians, and he narrowly defeated Richard Nixon — Eisenhower's Republican Vice President — in the 1960 presidential election.

Sputnik 1, the first man-made object to enter space, was launched from the Baikonur Cosmodrome in Kazakhstan on October 4, 1957. Its name means "fellow traveler", and the world thrilled to a new tune – Sputnik's famous signal "beep", picked up all around the world as Sputnik orbited. This picture shows a replica recently exhibited at the NASA *A Human Adventure* touring exhibition.

The seeds of the Apollo program dated back to 1960, during the Eisenhower Administration, when space planners envisioned it as a follow-up to the Mercury program. But Eisenhower was ambivalent about manned spaceflight. As a US Senator prior to his presidential bid, Kennedy was actually opposed to the space program, suggesting its termination. But as he took the presidential reins, politics changed the parameters of the race, and things got off to a slow start. Kennedy retained Eisenhower's science advisor Jerome Wiesner, who was actively opposed to manned spaceflight. Moreover, the new President was turned down by 17 nominees for the position of NASA Administrator before James E. Webb accepted the post. Webb was an experienced Washingtonian who had served President Harry Truman as Undersecretary of State.

Kennedy also established a so-called National Aeronautics and Space Council, and appointed Vice President Lyndon Johnson, from Texas, home of great interest in space and a burgeoning space industry, to lead this group. Johnson was experienced with NASA from his time in the Senate, as he helped to create the agency. Several months before his May 25 speech, Kennedy actually proposed international cooperation in space, during his annual January State of the Union Message. But Soviet Premier

Nikita Khrushchev rejected this idea, not wanting to expose the details of the Soviet programs in rocketry and space exploration. This caused Kennedy to accept the idea of pushing ahead with an independent US program, enticed him into the idea of shooting for the Moon, and accelerated the space race as a competitive political arena.

Meanwhile, in the Soviet Union, space policy had already been going full speed ahead. Soviet Premier Khrushchev, now 67 years old, fiery and completely self-assured, had a highly accomplished political career behind him. He was also First Secretary of the Communist Party, and had been celebrated for de-Stalinizing the Soviet Union and for pushing the Soviet space program forward. Born in the village of Kalinovka, near the present-day border of Russia and Ukraine, Khrushchev started off as a metal worker but became a political commissar during the Russian Civil War. Working his way upward in Russian politics, assisted by one of Joseph Stalin's associates, Lazar Kaganovich, Khrushchev took governance of Ukraine in 1938. During the Second World War — to Soviets the Great Patriotic War — Khrushchev was a commissar, serving as an aide to communications between Stalin and the leading field commanders. Following the German surrender, he

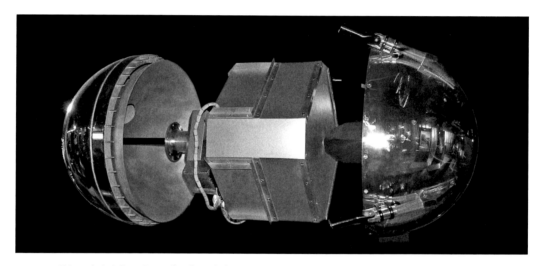

This exploded view of Sputnik 1 shows, left to right: the rear hemispherical heat shield, inside of which is a pressurized sphere; the cooling fan (small rectangular silver unit); the silver-zinc batteries to provide power for 22 days (in the large silver unit with chamfered corners); the transmitter (in the square black casing) which made Sputnik's distinctive beep, and the front hemispherical heat shield with the sockets for the antennas.

returned to Ukraine and then back to Moscow to advise Stalin on various matters.

Stalin died in 1953, and this created a power vacuum in the Soviet Union. In early 1956, at a party congress, Khrushchev delivered the "secret speech" in which he denounced Stalin's purges — which he had supported when they happened — helping to create a less repressive atmosphere. The Soviet Premier ordered military cuts, tried to improve the lives of ordinary citizens, and in 1959 famously debated Richard Nixon over the two countries' political systems. He traveled to the United States, visiting Eisenhower, and left with relatively good feelings about the two nations.

All the while, the Soviet Union had pursued a complex, aggressive, and innovative rocketry and space program. It was a military program and was classified, and had its origins in the 1930s. Among numerous Soviet accomplishments were the first intercontinental ballistic missile, the first Earth-orbiting satellite (Sputnik 1), the first animal sent into space (the dog Laika, in 1957), and the first human sent into space and in Earth orbit (Yuri Gagarin). And the Soviet program had only just begun. The Soviets, too, had aspirations of heading to the Moon.

The origins of the program commenced with the great engineer Konstantin Tsiolkovsky, who produced forward-looking studies in the late 19th century and the

Russian rocket engineers and enthusiasts pose with their first liquid-fueled rocket, GIRD-X, in 1933. At far left, with a Red Army cap, is Sergei Korolev, who would lead the Soviet space program during the 1950s and 1960s.

Profoundly deaf, one of rocket pioneer Konstantin Tsiolkovsky's many inventions was an ear trumpet for his own use. Tsiolkovsky was photographed in his office on March 15, 1930.

early years of the 20th century. In 1929, Tsiolkovsky introduced the concept of the multi-stage rocket, a major factor in later plans to explore space. In Moscow, in 1931, the Soviet Union founded the Group for the Study of Reactive Motion (GIRD), a think tank of brilliant engineers and scientists that constituted the world's first professional rocketry program. With Tsiolkovsky near the end of his life, the young engineer Sergei Korolev was appointed as first director of GIRD and chairman of its technical council, and it was he who would go on to lead the development of the space program in the 1960s.

Korolev dreamed of using rockets to one day travel to Mars. In 1933, along with many coworkers, including the German-Russian engineer Friedrich Zander, Korolev and his group produced the first Soviet liquid-fueled rocket. Throughout the 1930s Soviet rocket technology rivaled that of Germany, but Joseph Stalin's purges removed much of the talent pool. Korolev himself was detained in the Gulag. While the Nazis terrorized the UK with the V-1 "Doodlebug" (also known as the Buzz bomb) and produced the more sophisticated, alarming V-2, the Soviets mastered for military use the Katyusha multiple rocket launcher, employed in full by the middle of the Second World War.

At war's end, the Russian engineers who inspected captured German rockets were astonished at the technical complexity and sophistication of the German rocketry program. Following the war, as political anxieties produced the climate for a cold war between the United States and Soviet Union, the United States moved a large number of ex-German scientists and engineers, along with 100 V-2 rockets, to the West. The scientists moving to America included Wernher von Braun.

For better or worse, the space race was now officially on. Both superpower countries were forging ahead, each determined to be the first to get to the Moon. In the United States, Congress

Sergei Korolev, brilliant father of the Soviet space program, is seen here at his command post having two conversations at the same time! This stereo view was created by pairing screen shots from a 1950s public information film, showing the testing of the new Russian space technology.

An informal view of Yuri Gagarin (left, in uniform) talking to Sergei Korolev, c.1961.

Sergei Korolev shakes hands with Yuri Gagarin, seen here from behind in full space suit. Korolev was never referred to by his real name; only as "The Chief Designer", until his death in 1966. He personally chose Gagarin to be the first man in space.

substantially increased funding for space research. James Webb began a significant reorganization of NASA, and the agency increased its levels of staffing across the board. Webb pushed quickly on the building of two major new centers: a launch operations center for the proposed large Moon rocket, and a manned spacecraft center to be used as a central control hub. After a great deal of discussion on all fronts, the launch center was provisionally planned for an area in Florida northwest of Cape Canaveral Air Force Station, a military center on the eastern coast of Florida near the small town of Cocoa Beach. Planners decided to build the Manned Spacecraft Center on land donated from Humble Oil and Refining Company, through Rice University, in Houston, Texas.

Because of the gift, Kennedy traveled to Houston and delivered another landmark speech on the space program on September 12, 1962, at Rice University. To a large crowd that included Webb, the Governor of Texas, and various senators and representatives, Kennedy again laid out the justification for lunar exploration.

"We set sail on this new sea because there is new knowledge to be gained," said the young President, "and new rights to be won, and they must be won and used for the progress of all people. For space science, like nuclear science and all technology, has no conscience of its own. Whether it will become a force for

This is the stray dog Laika who became the first animal in space, launched into orbit by the Russians on November 3, 1957, just a month after Sputnik 1, to prepare the way for human voyagers. Compassionate voices of protest were raised all over the world since there was no possibility of a safe return to Earth; she died imprisoned in her Sputnik 2 capsule.

good or ill depends on man, and only if the United States occupies a position of pre-eminence can we help decide whether this new ocean will be a sea of peace or a new terrifying theater of war. I do not say that we should or will go unprotected against the hostile misuse of space any more than we go unprotected against the hostile use of land or sea, but I do say that space can be explored and mastered without feeding the fires of war, without repeating the mistakes that man has made in extending his writ around this globe of ours."

"There is no strife," continued Kennedy, "no prejudice, no national conflict in outer space as yet. Its hazards are hostile to us all. Its conquest deserves the best of all mankind, and its opportunity for peaceful cooperation may never come again. But why, some say, the Moon? Why choose this as our goal? And they may well ask, why climb the highest mountain? Why, 35 years ago, fly the Atlantic? Why does Rice play Texas?"

"We choose to go the Moon!," said the President. "We choose to go to the Moon in this decade and do the other things, not because they are easy, but because they are hard; because that goal will serve to organize and measure the best of our energies and skills, because that challenge is one that we are willing to accept, one we are unwilling to postpone, and one we intend to win."

The speech again widely inspired Americans and the world, and Kennedy galvanized the listeners by condensing human history into 50 years, metaphorically saying "only last week did we develop penicillin and television and nuclear power, and now if America's new spacecraft [Mariner 2] succeeds in reaching Venus, we will have literally reached the stars before midnight tonight." Kennedy had defined a new paradigm in technological prowess, and it meant getting to the Moon.

Meanwhile, the whole drama of the unfolding space efforts took place against a terribly strained political backdrop. Following the Bay of Pigs invasion into Cuba the previous year, and as a response to the presence of American ballistic missiles in Italy and Turkey, Soviet Premier Khrushchev accepted Cuba's request to place nuclear missiles on the island in order to deter a potential invasion. In July 1962, Khrushchev met with Fidel Castro and construction of missile launch facilities in Cuba commenced.

A close-up of President Kennedy addressing the crowd at Rice University on September 12, 1962.

In the late fall of 1962, several weeks after Kennedy's Rice University speech, the world came closer to a potential nuclear war than virtually anyone knew at the time. In the so-called Cuban Missile Crisis, Kennedy confronted Khrushchev, and a tense, 13-day standoff began that would last from October 16 until its resolution on the 28th. The two superpowers, each wanting to race to the Moon, having failed minor advances toward cooperating on space programs, having cooperated 20 years earlier to save the world from its darkest forces, now stared at each other across a dark abyss.

Elections were underway in the United States and Kennedy had to ward off accusations that he was ignoring the Missile Crisis to help politics. American U-2 spy planes confirmed the existence of medium-range and intermediate-range ballistic missile facilities. At astonishing risk, the US Navy established a blockade, preventing further missiles from reaching Cuba. Kennedy announced that he would not allow more weapons to reach Cuba and demanded the missiles already there be disassembled and returned to the Soviet Union.

The Cuban Missile Crisis. U-2 spyplane photograph of a missile base at San Cristobel on Cuba.

After days of perilous standoff and fragile negotiations, Kennedy's demands were met. Khrushchev proclaimed that the missiles would be dismantled and returned to the Soviet Union in exchange for a declaration that Cuba would not be invaded again by the Americans. Kennedy secretly also agreed to dismantle the American missiles that had been deployed in Turkey.

The end of 1962 marked a watershed moment in the race for space. Tensions between the United States and the Soviet Union reached a breaking point and then gradually began to subside. Both countries were committed to reaching out for the Moon, to winning the show of political dominance, to becoming the first nation to reach another world, to walk on its surface.

Someone, in the years to come, would make a giant leap for the human race. No one as yet knew who it would be or how or when it would happen. But the powers of human beings had been focused in a new way, with a new goal: to reach another world and to see how that effort would transform life on Earth. The heroic struggle that would take place over the years to come would astonish the participants.

In one sense, humanity – by its sheer commitment to this mutual goal – had already taken one small step.

A US Navy P-2H Neptune flying over a Soviet cargo ship with crated Il-28 missiles on deck during the Cuban Missile Crisis.

2 THE RACE HEATS UP
(MERCURY, VOSTOK 6)

Before John F. Kennedy challenged Americans to go to the Moon, the goals of the American space program had been somewhat nebulous. Maybe there would be circumlunar flights; maybe the nation would forge a program built around an orbiting space station. Kennedy's vision forged a sharp new focus and created a scramble within NASA to prepare for a manned lunar mission. At the time, Alan Shepard was the only American to have flown into space, and no American had yet orbited Earth. Many in the scientific community were skeptical about the nation's ability to get to the Moon before the 1960s ended. One thing was clear: a massive expansion of the space agency was necessary, and this would have to be done post-haste.

Within NASA, earlier feasibility studies by various companies were scrapped. The major new lunar exploration program, named Apollo for the Greek god of light, music, and the Sun, would take a new and more serious approach. The agency conducted internal engineering studies overseen by Belize-born mechanical engineer Max Faget, the designer of the Mercury spacecraft, who proposed a design consisting of a command and service module pairing. In late 1961, North American Aviation won the contract to build the spacecraft. Behind the scenes, even before Kennedy's two famous speeches, the wheels of change had been turning at the space agency. In the summer of 1960 NASA had established the Marshall Space Flight Center on

Astronaut Alan Shepard, the first American to fly in space, on the deck of the USS *Lake Champlain* after the recovery of his Mercury capsule, *Freedom 7*, in the western Atlantic Ocean, May 5, 1961. And on the opposite page, photographed in the *Freedom 7* capsule.

In 1959, engineers Charles Donlan, Robert Gilruth, and Max Faget inspect a Mercury capsule model.

the Redstone Arsenal, near Huntsville, Alabama. Here the heavy-lift launch rockets would be designed and built for the Apollo program, and these would come to be known as Saturn launch vehicles.

With the realization that NASA would need a new facility for spacecraft operations, the government commenced building the Manned Spacecraft Center on the gifted land near Houston. Previously, some spacecraft operations had been handled from the Langley Research Center near Hampton, Virginia. Other facilities, like the control

Dr Wernher von Braun, the father of rocket technology in the United States, from an early undated NASA portrait.

center of the Mercury missions at Cape Canaveral Air Force Station in Florida, were deemed to be too small for future growth. The ability to manage maneuverable spacecraft like those planned for Apollo would require a far larger and much more sophisticated control operations center than the ones located at Cape Canaveral, Langley, or at the Goddard Space Flight Center in Maryland.

By late 1962, Robert Gilruth, the aerospace engineer in charge of spacecraft operations, moved his group to Houston, although he initially was "aghast" at the planned lunar landing and its timescale. Aerospace engineer Christopher Kraft, NASA's first flight director, also took on a substantial role, and he would become the most significant influence on the Mission Control Center as the work on it unfolded.

Construction of the new Mission Control Center began in 1963. The site comprised 1,620 acres, and the new Mission Control Center was contained within building 30. This building held two rooms such that one could be used for practices and training, and the other as the operations center for live missions taking place in real time. Eventually, the center would be renamed the Christopher C. Kraft, Jr. Mission Control Center, in recognition of Kraft's unique service as flight director, and the entire facility was renamed the Johnson Space Center in 1973, following Lyndon Johnson's death.

The launch of a V-2 rocket in 1943, taken four seconds after liftoff from Peenemünde on the Baltic Sea island of Usedom.

The other substantial building activity required for the Apollo program, beyond Mission Control in Houston and the Marshall Center in Alabama, was the vast expansion of the Cape Canaveral facilities in Florida. This growth was linked to the Saturn rocket program, started as a response to the US Department of Defense's call for a heavy-lift vehicle to carry military and communications satellites into orbit. The American launchers of the day were wholly inadequate, able to lift only about 1,400 kilograms into orbit, while the new requirement would be 9,000 to 18,000 kilograms. So the engineering team led by Wernher von Braun, German-American aerospace engineer who had worked on Hitler's V-2 rockets during the Second World War, attacked the problem. In the late 1950s, before the call for lunar exploration, von Braun and his team began working on this challenge.

The Marshall team led by von Braun calculated that the new rocket would require a lower-stage booster with a thrust of about 1.5 million pound-force at takeoff. At first the von Braun group proposed a cluster of largely existing rockets, including Jupiter, Redstone, Atlas, and Titan components. After NASA entered the picture, the von Braun team tried to collect a multitude of heavy-launch options involving

Mercury *Friendship 7* replica on display at NASA's *A Human Adventure* exhibition in Milan.

several agencies and branches of the military. Recognizing that the US program had been confused and mishandled, and that the Soviet program was believed to be far ahead, von Braun produced a substantial report by early 1959, and NASA selected his proposal. Test firings for what became the Saturn boosters began in 1959. That summer, Department of Defense officials nearly canceled the program, but then changed their minds.

By 1961, von Braun's team had pushed ahead and achieved a successful, maiden test flight of the Saturn I rocket. Throughout the next 3½ years more test flights took place with Saturn rockets, giving von Braun and his crew of engineers the confidence to begin to evolve the design toward what would actually be used several years later during the active Apollo program.

As the Apollo program worked through its developmental infancy, the United States continued Project Mercury, following on from Alan Shepard's suborbital flight. In July 1961 Virgil "Gus" Grissom became the second American in space, with his 15-minute flight in *Liberty Bell 7*, infamous for the loss of the capsule on splashdown when the explosive hatch blew. (The capsule would be recovered from the ocean

John Glenn pictured prior to his historic flight on February 20, 1962.

Gus Grissom entering his F-106 Delta Dart.

floor in 1999.) On February 20, 1962, John Glenn became the first American to orbit Earth in his *Friendship 7*, during a flight that lasted nearly five hours and made three orbits. Three more Mercury astronauts followed: Scott Carpenter with a flight on May 24, 1962; Wally Schirra with his nine-hour mission on October 3, 1962, and Gordon Cooper with his flight of more than one day and ten hours, completing 22 orbits, on May 15, 1963.

While the Americans scrambled to catch up, driven by the increasingly intense cold-war competition, the Soviet Union sprang ahead with continued successes in their space program. In all respects, the Russians continued to lead the space race, and they had clear and growing intentions of landing a man on the Moon — before the Americans.

The Russian program also benefited enormously from captured German engineers and scientists, who went to work for the Soviet Union, along with numerous captured plans and engineering drawings. Coordinating the new Soviet missile program was Dmitry Ustinov, who worked with Korolev, released from prison, to reestablish a Soviet rocketry program throughout the 1950s. One of von Braun's managers in the

Valentina Tereshkova, the first woman to fly in space, is pictured here at a reception at the Piccadilly Hotel, London, where she was presented with a gold medal from the British Interplanetary Society.

The Vostok 6 capsule in which Valentina Tereshkova made her historic flight. Inside the capsule she faced the grey control panel at top center of this photograph.

V-2 program, Helmut Gröttrup, also worked extensively on the new Soviet program, creating a replica rocket like the V-2 and liquid-fueled rockets, which led to the R-7 Semyorka, the Soviet intercontinental ballistic missile tested in 1957.

The fuel that pushed the Soviets and Americans to spin up a race for space really spilled over in the 1950s. Driven by the military, cloaked in uncertainty and worry over which of the two superpowers would seem the most powerful, the programs lunged forward. The R-7 could not only deliver nuclear warheads but could also support space launches. Following Sputnik 1 in 1957, Korolev finally received the instruction he had long waited for — to accelerate the manned space exploration program. The team produced the Vostok spacecraft which was used by Yuri Gagarin for his historic first spaceflight. Now driven by his dream to send humans to Mars, for a time Korolev thought far ahead of the American goal of landing on the Moon. He thought of the Red Planet, which could be visited by closed-loop life-support systems and electrical spacecraft engines launched from a space station orbiting Earth.

At first, in fact, Nikita Khrushchev was not enthusiastic about competing with what became the Apollo program. The Soviet Premier had a good relationship with Korolev and the rocketry and engineering teams, but rather than ordering one project at a time for propaganda purposes, Khrushchev was most interested in the military capability of the missile program.

Nevertheless, the Vostok program moved forward in tandem with the American Mercury program. Following Gagarin's flight on April 12, 1961, on August 6 of that year, Gherman Titov became the second human to orbit Earth, and spent the first full day in space, in Vostok 2. A year later the Vostok program produced another amazing first when Vostok 3 and Vostok 4 flew into space simultaneously, with Andriyan Nikolayev and Pavel Popovich piloting the two spacecraft, with Popovich later saying at a news conference that he saw Vostok 3, the other craft, "once" during the flight and that "it looked like a very small moon in the distance."

The Vostok program produced one more joint flight, which was amazing and very unusual, this time in June 1963. Following the last of the Mercury flights a month earlier, Vostok 5 and Vostok 6 featured Valery Bykovsky as pilot and another historical first — the first woman in space, Valentina Tereshkova, piloting Vostok 6. The craft launched on June 14 and June 16 and after nearly 5 days for Vostok 5 and almost 3 days for Vostok 6, landed on June 19. Again, with the successful coordinated flight and the first woman in orbit, the world was amazed and transfixed in wonder by the Soviet space program.

The Soviet launch and operations facilities, meanwhile, had also been a longstanding and complex project, originating in the 1950s. As far back as 1955, the Soviet government laid plans for what they termed Scientific Research Test Range No. 5, which would serve as a test center for that first intercontinental ballistic missile, the R-7 Semyorka. The site commission was led by General Vasily Voznyuk, and consisted of a variety of military and engineering minds that included the great Sergei Korolev. The commission wanted to settle the facility in an area away from populated regions and built within a large open plain, which would decrease dangers to civilians and make radio transmissions from distant relay stations receivable.

With all of the considerations in mind, the commission chose a site near the village of Tyuratam, in the Kazakh Steppe, in what is now Kazakhstan. The isolated location, more than 4,000 kilometers (2,500 miles) from Moscow, made the project one of the most expensive of that era for the Soviet Union. Launch facilities and control and operations centers had to be constructed, but also hundreds of kilometers of roadways built. As with similar isolated research facilities, like those of the Manhattan Project, the Soviet Union built an entire support city around the facility, with housing, schools, and everything else the workers would need. Later, in 1966, the town would be named Leninsk, and much later, in 1995, its name changed to Baikonur.

But the launch facility's name was Baikonur from the start — more specifically, the Baikonur Cosmodrome. The name is rumored to have been a deliberate attempt to misdirect Western attention to a small town called Baikonur 320 kilometers (199 miles) northeast of the launch center. By the time of its construction in the 1950s, Baikonur was the world's first spaceport, and it is now immense in size, measuring an ellipsoidal 90 by 85 kilometers (56 by 53 miles) in extent. Much later, the launch pad at Baikonur used for both Sputnik 1 and for Yuri Gagarin's first manned space flight was renamed Gagarin's Start to honor the first cosmonaut.

The Americans had one thing going for them, however. NASA Administrator James

Valentina Tereshkova, former factory worker and skydiver, became the first woman in space when she flew in Vostok 6 on June 16, 1963.

Webb was in charge of the agency's programs throughout most of the 1960s. In the Soviet Union, the architecture featured several competing design groups, which sometimes led to inefficiencies. With the end of the 1950s and the beginning of the 1960s, Korolev's design group had some competition from other engineers and their comrades — specifically from Vladimir Chelomei, who had invented the first Soviet pulse jet engine; Valentin Glushko, who had studied the German rocketry program; and Mikhail Yangel, a pioneer in designing intercontinental ballistic missiles.

With the support of Khrushchev, Chelomei was assigned the development of a rocket that could propel humans around the Moon and to produce rockets that could support a military space station. But he was relatively limited in experience with space missions, and so he developed his ideas slowly. Glushko did not always get along well with Korolev, and did not participate in building the heavy boosters that Korolev needed for a manned trip to the Moon. Yangel had been an assistant of Korolev's and started strongly on his own experiments. But in 1960 a terrible accident slowed his participation. At Baikonur, on October 24, 1960, Yangel, along with Chief Marshal of Artillery Mitrofan Nedelin and many others were conducting a test of a prototype Intercontinental Ballistic Missile (ICBM) called the R-16. The second stage engines fired prematurely, killing Nedelin and 78 military and technical personnel in the area. (Some reports place the dead at as many as 126.) The accident, which came to be known as the "Nedelin Catastrophe," was a stunning setback. Yangel survived only because he had walked off to smoke a cigarette, and he later suffered a heart attack that slowed his work for years.

The American progress on the Apollo program surprised and alarmed the Soviets. In response to Apollo, each of the several principal engineers pushed for their own programs and their own engineering groups. But Korolev used his influence to move forward with a lunar rocket design group, and by 1964 it would formally declare its intentions to go to the Moon, some three years after the Americans had. By then, Korolev was the unrivaled chief of the design group. But a great deal of time and effort had been wasted on multiple ideas, projects, competing designs, and rivalries. Korolev stated that the Soviets should land a man on the Moon by 1967, and he believed this would firmly beat the Americans. The space race was now a race to take steps on another world, a movement that would result in the giant leap for humanity, regardless of who arrived first and exactly how they got there.

Before even the year 1963 was out, however, another event stunned the world. In November of that year John F. Kennedy traveled to Texas to smooth over frictions in the Democratic Party. He sought to bring together liberal Ralph Yarborough and conservative John Connally, and was determined to raise funds for the party and begin his quest for reelection the following year.

Dallas was an ultra conservative city, hostile to Democrats and filled with dissatisfaction over the Massachusetts liberal and his policies. Despite concerns over anti-Kennedy feelings there, the President traveled to the city with his wife Jackie and arrived at Love Field on November 22. The President's parade route toward a speech at the Dallas Trade Mart had been made public several days beforehand. The day began brightly, and the President and First Lady waved and smiled and greeted countless thousands.

When the presidential motorcade turned onto Elm Street in front of the Texas School Book Depository at 12:30 p.m. CST, three shots rang out. A sick, angry young man, Lee Harvey Oswald, shot and killed Kennedy, hitting him once in the upper back and neck, and a second time, fatally, shattering his skull. The American nation mourned, the world was stunned, and for a time it seemed that Kennedy's dream of landing a man on the Moon had faded into a murky, dark background of disaster.

A nighttime photo shows gantry arms opening on a modern Soyuz rocket at the Baikonur Cosmodrome.

Gemini IV crew members Ed White (Pilot), Jim McDivitt (Command Pilot), and backup crew Frank Borman and Jim Lovell pose with a model of the spacecraft, September 10, 1964.

January 15, 1966 — Astronauts Neil Armstrong (center), Command Pilot, and Dave Scott (right), pilot of the Gemini VIII prime crew, smiling in their silver spacesuits during water egress training aboard the NASA Motor Vessel Retriever in the Gulf of Mexico. At left is Dr Kenneth Beers, MD.

3 THE PLAYERS
(GEMINI, VOSKHOD)

In the immediate aftermath of the Kennedy assassination, Vice President Lyndon Baines Johnson was sworn in as the 36th President of the United States. As Chairman of the National Aeronautics Space Council, Johnson had been instrumental in pushing the American space program forward. Johnson became President just two hours and eight minutes after the assassination, taking office on Air Force One on the tarmac in Dallas, the oath administered by Judge Sarah Hughes. As the nation's leader, Johnson kept the initiatives for the aggressive American space program rolling ahead at full speed.

For the Mercury program, the first American astronaut group, the so-called Mercury Seven, consisted of Scott Carpenter, Gordon Cooper, John Glenn, Gus Grissom, Walter "Wally" Schirra, Alan Shepard, and Donald "Deke" Slayton. In 1962, NASA anticipated the next stage of American exploration by choosing astronaut group 2, also known as the New Nine. They were Neil Armstrong, Frank Borman, Charles "Pete" Conrad, James "Jim" Lovell, James McDivitt, Elliot See, Thomas Stafford, Edward "Ed" White, and John Young. In order to start up multiple, overlapping programs, hoping to catch up with the impressive performance of the Soviets, in 1963 NASA also selected a third astronaut group. The members were Edwin "Buzz" Aldrin, William "Bill" Anders, Charles Bassett II, Alan Bean, Eugene "Gene" Cernan, Roger Chaffee, Michael Collins, Walter "Walt" Cunningham, Donn Eisele, Theodore Freeman, Richard "Dick" Gordon, Russell "Rusty" Schweickart, David "Dave" Scott, and Clifton Williams.

As the Apollo program slowly moved forward, an ambitious movement commenced that would lay the groundwork for Apollo and test many of the flight capabilities required for landing on the Moon. Designed to follow Mercury, Project Gemini commenced its initial stages in 1961 and concluded five years later. The project was far more sophisticated in terms of spaceflight than Mercury, utilizing a two-man capsule that would allow proving of concepts, coordination, and timing that would lead to landing a man on the Moon.

Gemini's chief designer was Jim Chamberlin, a Canadian-born aerospace engineer who had joined NASA in 1959 following the cancelation of Canadian programs. Chamberlin pushed some key concepts, such as pairing the two-man spacecraft with a "bug" that could be detached and land on the lunar surface. For the Gemini program, this idea was rejected, but the concept led to an acceptance of a lunar orbit rendezvous for the Apollo program to follow. Hailed as a brilliant engineer by those around him, Chamberlin left NASA by 1970, but not before leaving a huge imprint on the agency with his primary contributions to the Gemini program.

Gemini's objectives were simple: NASA wanted to test the endurance of humans in space over relatively long periods, as it might take eight days to two weeks for a lunar mission; to practice extravehicular activities (spacewalks); to practice spacecraft maneuvers between two systems, with control and docking procedures; and they wanted to practice reentry into Earth's atmosphere and achieve successful landings.

In 1962, the Titan II missile had replaced the Atlas as the US Air Force's second-generation ICBM. It would serve as the primary launch vehicle for Gemini. The two-man spacecraft was built by McDonnell Aircraft, the same contractor that built the Mercury capsules. It was an enlarged version of Mercury, but had retrorockets, propulsion, oxygen, and water in a detachable module behind the capsule, which was called the Reentry Module. NASA now modernized their craft into a modular system such that components could be replaced with relative ease. The Gemini craft was also the first to employ an onboard computer.

As always, with the Gemini program safety was a top concern. Pushing the boundaries of exploration is always a dangerous business, and so the Gemini engineers built an ejection seat mechanism into the Gemini system rather than the older escape tower system powered by a solid-fuel rocket. Ejection seats, the engineers determined, could separate the astronauts from a malfunctioning rocket. Max Faget, the Mercury engineer, was not enthusiastic about this turn, believing the window for escape would be too narrow. Later, he simply said, "The best thing about Gemini is that they never had to make an escape." Gemini also incorporated some useful tools from the aviation industry, including in-flight radar and an artificial horizon.

Gemini flight planners originally intended that the craft would come down and make a landing on solid ground. The idea was to use a flexible wing rather than a parachute to achieve a soft landing. But the complex attachments of the wing to the capsule ultimately meant that designers dropped the notion and returned to a conventional parachute approach into the ocean, with ships and helicopters awaiting the craft and crew.

The first unmanned test flight of the Gemini program, Gemini I, took place on April 8, 1964, and lasted four hours and 50 minutes. Crew life-support systems were replaced with ballast, and engineers simply studied the performance of the craft during flight. Four large holes drilled into the craft's heat shield ensured that the capsule would disintegrate on reentry into the atmosphere. Following three orbits, Gemini I fell back to Earth and burned up, its mission a success.

Further steps with the Gemini program took place the following winter, when on January 19, 1965, engineers launched Gemini II, another unmanned test flight. The primary objective with this mission was to test the craft's heat shield. The mission lasted a grand total of 18 minutes and 16 seconds, and the capsule came down in the Atlantic Ocean, recovered by crew of the USS *Lake Champlain*. A far more critical point would arrive with Gemini III, on March 23, 1965, the first manned

The first manned Gemini flight took place on March 23, 1965, when Gus Grissom and John Young experienced nearly five hours in orbit. The capsule was nicknamed *Molly Brown* after the Broadway musical and was a reference to Grissom's Mercury craft, *Liberty Bell 7*, which sank shortly after splashdown.

Gemini mission. The first Gemini crew consisted of Gus Grissom and John Young, and they nicknamed their capsule *Molly Brown* — Grissom's humorous reference to *The Unsinkable Molly Brown*, following the loss of his Mercury capsule at sea. The mission lasted nearly five hours and gave mission controllers their first test of the Gemini capsule as operated by humans, provided a first in the orbital maneuver by a manned spacecraft, and made NASA officials feel good about the direction of the Gemini program.

However, the Americans, as usual, were not the only game in town, and the Soviets continued to upstage them with regularity. The second Soviet human spaceflight project, Voskhod (Russian for "ascent," or "dawn"), pushed onward at high speed in the wake of the successful Vostok program. Engineers developed a new craft, Voskhod, which was essentially a Vostok craft with a second, solid-fuel retrorocket added to the descent module. To accommodate the added mass, the Soviets employed a more robust launch vehicle for this phase, the 11A57, which was a Molniya 8K78M

 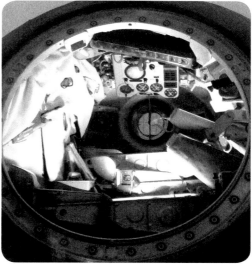

A view inside the Voskhod 1 capsule. Near the center you can see the porthole, and above that is the main control panel.

The capsule of Voskhod 1 after landing. Voskhod 1 was the first spaceflight with more than one crewmember, and the first in which the crew did not wear spacesuits. It was launched on October 12, 1964 carrying cosmonauts Vladimir Komarov, Konstantin Feoktistov and Boris Yegorov and made 16 orbits of Earth.

Voskhod in launch shroud.

rocket, a system that later became the core of the Soyuz rocket booster.

The Voskhod capsule could carry as many as three astronauts, as opposed to Gemini's two, and the seats were arrayed in a 90 degree angle relative to those of the Vostok craft. This meant that the crew had to move their heads around to see the instruments with sufficient precision. With air-cooled avionics and environmental systems, the craft required an inflatable airlock that was jettisoned after use. Engineers also added a solid-fuel-powered braking rocket which fired moments before touchdown to cushion the impact. The new craft lacked ejector seats for the crew, such that escape from a major malfunction would not be possible.

In October 1964, Soviet engineers first tested the Voskhod system in an unmanned flight designated Kosmos 47. Although without crew, this was the first test flight of a spacecraft designed for a multiple-person crew in history. From Baikonur, the craft launched and orbited for almost exactly one day before landing.

Just six days later, on October 12, 1964, the Soviet space program marked a major first in the history of space exploration. Voskhod 1 launched from Baikonur carrying a crew of three cosmonauts, making it the first mission in history to carry multiple human beings. It was also the first spaceflight in world history to occur with crewmembers not using spacesuits — the capsule, with three inside, lacked room for them. The flight set a new manned spacecraft record for altitude, with the craft orbiting at 336 kilometers (209 miles). The crew also consisted not simply of test pilots, but for the first time ever, a combination of pilots and scientists.

Voskhod 1's team of cosmonauts included the Command Pilot Vladimir Komarov, who was a test pilot and engineer. Born in Moscow, Komarov was an ex-Air Force pilot who was regarded as one of the most highly-qualified cosmonauts in the Soviet program. At first he did not medically pass muster to become a cosmonaut, but his hard work and perseverance overturned that situation. He was among a large number of cosmonauts who lived and trained at the military base where training took place, which eventually became known as the Yuri A. Gagarin State Scientific Research-and-Testing Cosmonaut Training Center, in Moscow Oblast. Journalists eventually found out about the originally secret facility and began to call the center "Star City."

Komarov's companions in the small Voskhod capsule were the engineer Konstantin Feoktistov and, for the first time in space, a medical doctor, Boris Yegorov. The three cosmonauts orbited Earth for just a shade more than one day 17 minutes, and the mission was tense given the fact that if any major mishaps occurred, there would be no way to rescue the travelers. The cramped conditions inside the capsule led to the notion of keeping the mission short. During the flight, Nikita Khrushchev spoke to the crew via a telephone link. Soon after the conversation, he was called to Moscow, and amazingly enough, learned that he was to be removed from office and expelled from the Communist Party. The crew returned to Earth on October 13 and were greeted by Leonid Brezhnev and Alexei Kosygin, making their first public appearance as the Soviet Union's new leaders.

The flight of Voskhod 1 took place before the first of the American Gemini flights. And there was more to come from the program. On March 18, 1965, the Soviets launched their second Voskhod mission, Voskhod 2, which would also achieve much in the way of spaceflight history. A new and improved craft, the Voskhod 3KV, was used for this flight which would carry two spacesuited cosmonauts. They were Commander Pavel Belyayev, a fighter pilot with extensive experience on a wide variety of aircraft, and Alexei Leonov, an Air Force pilot.

Alexei Leonov is a talented artist, and during the flight of Voskhod 2 he became the first person to create a work of art in space.

The crew of Voskhod 1: left to right, Vladimir Komarov, Boris Yegorov, and Konstantin Feoktistov at the Baikonur Cosmodrome after the end of their mission.

Late in the morning, the craft was launched from Baikonur, and the crew deployed an inflatable airlock once in orbit. The mission lasted just over one day and two hours, and the craft's altitude varied between 167 kilometers (104 miles) and 475 kilometers (295 miles). About 90 minutes after launch, Leonov made spaceflight history when, in his specially crafted suit, he left the capsule and performed the first human spacewalk.

Leonov's spacewalk lasted 12 minutes and 9 seconds, during which time he attached a camera to the end of the airlock to record his activity. But such a first carries lots of unexpected issues: the still camera mounted on his chest was inoperable because the suit he wore ballooned out in volume and he couldn't reach the shutter release. More alarmingly, after he completed the spacewalk, Leonov found the suit had expanded so much that he couldn't reenter the airlock. Cool and determined, he released some of the suit's pressure to enable bending the suit's joints. But he passed the recommended safety limits, and did not report this fact, and the radio and television transmission of the walk ended in some confusion. Despite the alarming challenges, Leonov was able to squeeze back into the craft. Inside, Leonov and Belyayev were so cramped that they could not return to their seated position,

Alexei Leonov during preparations for his historic first spacewalk during the Voskhod 2 mission, smiling in his helmet, which was equipped with a microphone.

On March 18, 1965, during the Voskhod 2 mission, Alexei Leonov squeezed himself free of the craft's airlock and became the first human ever to float freely in space.

Alexei Leonov photographed during the historic spacewalk. The Soviets knew that NASA was also preparing an astronaut for a spacewalk and cabled the news to the world as soon as Leonov was back inside the spacecraft. A TV camera recorded the footage we have used to assemble these stereo views of the event.

throwing off the capsule's center of mass, which caused it to land way off course. Nonetheless, the two regained control and the craft made a safe landing using the manual mode, and history recorded a triumph from this first human spacewalk, which had been experimental and highly dangerous.

The Russians mounted just two Voskhod flights, but they were successful and victorious, each marking significant "firsts," and again they had upstaged the Americans. Gemini III, though a success, had its thunder stolen by the remarkable nature of Voskhod 2, which took place just days beforehand.

These spectacular firsts enabled the Soviets to maintain the upper hand. And more Voskhod flights were planned, at least four more missions, but delays in the Soviet program allowed Gemini flights to accomplish some of what Soviet mission planners had on the table. This period, although dominated by the Soviets, amounted to a washing back and forth in terms of momentum. In the end, the goals had been met, and the Americans for the most part bested; now the Soviets abandoned further Voskhods in order to focus on developing the next stage in their program, the Soyuz effort.

Launched in the mid-1960s, the Ranger fleet of spacecraft provided live television transmissions of the Moon from lunar orbit for the first time. These transmissions showed surface features as small as 25 centimeters (10 inches) across and provided more than 17,000 images of the lunar surface. The detailed photographs allowed scientists and engineers to study the Moon in greater detail than ever before, paving the way for the Apollo landings.

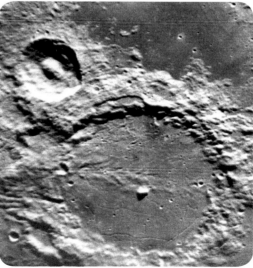

Just another photo of lunar craters? No! This stereo was constructed from Ranger 9's last glimpses of its destination, seconds before it crashed into the Moon. NASA's Ranger 9 probe carried TV cameras and the end of its journey was broadcast live, with millions watching across the United States. This stereo view shows the crater Alphonsus, site of the impact, and the much smaller crater Alpetragius.

Soyuz was secretly developed in the early 1960s as the Soviet mainline program to deliver a manned crew to the Moon. It followed Vostok and Voskhod, and employed a basic spacecraft design developed by Korolev. The Soyuz spacecraft was intended not only for a lunar mission, but also for many other projects that never saw the light of day. The Soyuz craft consisted of three main components: a spherical orbiting module, a small reentry module, and a cylindrical service module. The planned date for a lunar landing was 1967, to mark the 50th anniversary of the October Revolution, or, if need be, 1968. For a time the Soviet planners were confident that they could beat the Americans to the Moon. Leading cosmonaut candidates to make up the first lunar crew included Yuri Gagarin and Alexei Leonov.

This photograph of Ed White's first US spacewalk was taken by Commander James McDivitt over a cloud-covered Pacific Ocean. The maneuvering gun is visible in White's right hand. The visor of his helmet is gold-plated to protect him from the unfiltered rays of the Sun. When the time came to go back in, he is famously quoted as saying: "It's the saddest day of my life."

The American Gemini program, meanwhile, continued to speed along with success. Gemini IV, a four-day mission in June 1965, featured the first American spacewalk, by Ed White, while his fellow crewmember James McDivitt stayed within the capsule. In August 1965, Gordon Cooper and Pete Conrad broke an orbit duration record by

An artist's impression of the first piloted rendezvous in space, December 15, 1965. This was the very first time the US achieved a human spaceflight "first" – and this step was essential to develop the necessary skills to succeed with the Apollo Moon landing missions.

The stereo views on this page are of the first manned rendezvous in space of the two Earth-orbiting spacecraft, Gemini VI-A and Gemini VII. This one shows Gemini VI-A as observed and photographed from Gemini VII, when the two craft were about 12 meters (40 feet) apart.

These pictures show the opposite view – Gemini VII as it appeared from the hatch window of the Gemini VI-A capsule during rendezvous maneuvres. The two spacecraft did not dock, but maneuvered as close as one foot from each other. They flew in formation for more than five hours, orbiting Earth 3½ times.

spending nearly eight days in their Gemini V capsule. The next flight, Gemini VII, took place in December 1965 and was crewed by Frank Borman and Jim Lovell. This mission orbited Earth for nearly 14 days and made a total of 206 trips around the planet, investigating the effects of two weeks in space on the human body. The final Gemini flight of 1965 came with Gemini VI-A, whose crew of Wally Schirra and Thomas Stafford performed the first manned rendezvous with another spacecraft, Gemini VII. The rendezvous occurred on December 15, when the two craft approached within one foot of each other, and they easily could have docked had the craft been equipped for doing so. The Gemini VI-A crew broadcast a message suggesting they had spotted Santa Claus's sleigh, and then hammered out a harmonica rendition of "Jingle Bells," thus producing the first music created in space by humans.

The year 1966 brought an aggressive continuation of the Gemini program. In March, Gemini VIII spent more than 10 hours in orbit with the crew of Neil Armstrong and Dave Scott, and marked another historic first. The craft docked with an unmanned Agena Target Vehicle, establishing the first in-orbit docking of two spacecraft, which would be required for the Apollo missions. A glitch occurred while the two craft were docked, however: the Gemini craft experienced a thruster malfunction, which caused a calamitous tumbling of the craft. Armstrong took control and attempted to thwart the tumbling, and he maneuvered the craft toward Earth in the first emergency landing of such a space capsule. By the time Gemini VIII approached a splashdown, the tumbling had reached one revolution per second, threatening the astronauts' consciousness. Nevertheless, Armstrong controlled the descent well enough to splash down in the Pacific, south of Japan.

In May, Gemini IX-A carried its crew, Tom Stafford and Gene Cernan, through a nearly four-day mission for another planned docking maneuver. The original

This is the Gemini Agena Target Vehicle – the unmanned target of the very first successful docking maneuver in space, performed by Neil Armstrong and Dave Scott during the Gemini VIII mission. The stereo was assembled from pictures taken by Scott above the west coast of Mexico just prior to docking on March 16, 1966 – the Agena was at a distance of 14 meters (45 feet).

Astronauts Neil Armstrong and Dave Scott sit with their spacecraft hatches open while awaiting the arrival of the recovery ship, the USS *Leonard F. Mason*, after the successful completion of their Gemini VIII mission. They're assisted by rescue personnel. The overhead view shows the Gemini VIII spacecraft with the yellow flotation collar attached to stabilize it in choppy seas. The green marker dye is highly visible from the air and is used as a locating aid.

unmanned docking craft failed, however, and so they sped toward a rendezvous with an Augmented Target Docking Adaptor. The Gemini craft rendezvoused with the unmanned craft, but a malfunction disabled their docking procedure. Additionally, Cernan conducted a spacewalk, but due to a fogged-up visor, overheating, and fatigue, he could not perform a free-flying maneuver using a self-contained rocket pack, as the original plan had dictated.

Nonetheless, Gemini continued to deliver. In July 1966 Gemini X was launched with its crew of John Young and Michael Collins for a nearly-three-day mission. The flight tested the dangers of high-altitude radiation, docking with an Agena booster that lifted them into a high-elevation orbit. They also met the defunct Agena booster from the aborted Gemini VIII flight that had experienced the thruster malfunction. Thus, this mission achieved the first double rendezvous in space. In September, Gemini XI carried Pete Conrad and Dick Gordon through a mission lasting nearly three days. They performed the first-ever direct-ascent rendezvous with an Agena Target Vehicle, docking just 1½ hours after launch. This simulated a Lunar Module (LM) docking with the Command Module during a future Apollo Moon mission. Gordon also accomplished two spacewalks during the mission, and they achieved a very high-altitude orbit.

Buzz Aldrin was the first astronaut to capture a selfie in space, during the fight of Gemini XII in 1966.

Following the Gemini XII splashdown on November 15, 1966, astronauts Buzz Aldrin (left) and
Jim Lovell are welcomed aboard the recovery aircraft carrier, USS *Wasp*, after four days in space.

The final Gemini mission, crewed by Jim Lovell and Buzz Aldrin, was Gemini XII, and it took place in November 1966. The mission lasted nearly four days and contained the fifth rendezvous and fourth docking with an Agena Target Vehicle, preparing for the fine control that would be needed with the Apollo missions. Further, Buzz Aldrin conducted three spacewalks. The ability of astronauts to work extensively outside spacecraft and their successful docking maneuvers with other vehicles paved the way for Apollo. Now the Americans had made a substantial demonstration of where their program was headed, and at impressive velocity.

In 1965 and 1966 NASA further demonstrated its commitment to the Apollo program by choosing two more groups of astronauts. Astronaut group 4 consisted of scientists, for the first time, rather than test pilots. They were Owen Garriott, Edward Gibson, Duane Graveline, Joseph Kerwin, Curtis Michel, and Harrison "Jack" Schmitt. Astronaut group 5, chosen in 1966, included Vance Brand, John Bull, Gerald Carr, Charlie Duke, Joe Engle, Ronald Evans, Edward Givens, Fred Haise, James "Jim" Irwin, Don Lind, Jack Lousma, Kenneth Mattingly, Bruce McCandless, Edgar Mitchell, William Pogue, Stuart Roosa, John "Jack" Swigert, Paul Weitz, and Alfred "Al" Worden. NASA now had an entire village of people in astronaut training.

The Gemini XII patch shows the Roman numeral XII at 12 o'clock, with the Gemini spacecraft pointing to it, representing the fact that Gemini XII was the last flight of the Gemini program. With Apollo to follow, the ultimate objective – the Moon – is symbolized by the crescent on the left.

The many achievements and increasing complexity of the Gemini missions, along with internal competition and shifting goals within the Soviet program, stalled many of the Soviet plans. And then came an event that proved to be disastrous. Suddenly, on January 14, 1966, Sergei Korolev, the lead Soviet rocket engineer and driving spirit behind the Russian lunar program, died. He had previously suffered a heart attack and accumulated numerous health problems, some of which resulted from his days in detention. Following an operation, the details of which are reported in conflicting ways by various sources, he declined rapidly and expired. Now the Soviet program was without its leading light, just as the Americans were gearing up in a huge way for the Moon missions to come.

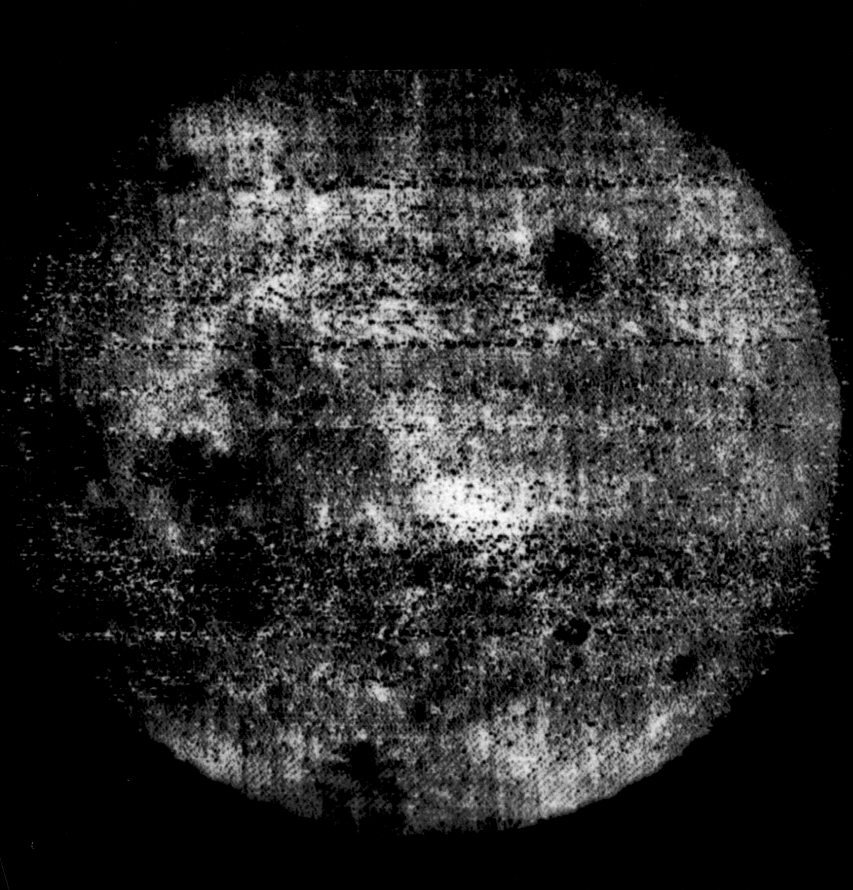

The very first image of the far side of the Moon, photographed by Luna 3 in 1959 from a distance of 61,700 kilometers (38,600 miles). The dark spot at upper right is Mare Moscoviense. The dark area at lower left is Mare Smythii, and the small dark spot at lower right with a white dot at its center is the crater Tsiolkovsky and its central peak.

4 THE FAR SIDE
(LUNA, APOLLO)

The years 1966, 1967, and 1968 would be pivotal ones for the history of the world. Unlike the first period of the cold war, the 1950s, which were marked by sleepy conservatism, the 1960s – and especially the middle 1960s – commenced a new era in politics and culture. The United States and the Soviet Union were locked in a cold-war competition, a space race, but in the West, especially, the times were changing. The Vietnam War was escalating, and in America antiwar sentiment was growing rapidly.

In 1964, the Beatles had burst onto the scene and begun to revolutionize popular music. In the summer of 1966, their album *Revolver* was released and reflected a rapid evolution of counterculture, recreational drug use, protesting the status quo, and experimenting with everything from fashion to films. The first glimmers of psychedelia had emerged. Even as a young child (born 1961), the son of a university professor in southern Ohio, I could sense the changes all around me.

The spirit of the sixties grew like a bonfire during 1966 and would only accelerate in the years to come. The rapidly escalating social change painted a backdrop against which the space race continued onward, although its nature would now dramatically change, even if all the participants didn't know it for a while. In the Soviet Union, the death of Korolev was a blow that would be very hard to recover from. Despite the great successes of the Voskhod program, the Soviets largely watched and developed their new program, Soyuz, as the Americans suddenly surged ahead with multiple successes in the Gemini program.

This small probe, Luna 3, was launched on October 4, 1959, two years to the day after the first great Soviet space triumph, the launch of Sputnik 1. It captured the first-ever photos of the far side of the Moon.

The Russian Luna program, utilizing unmanned spacecraft, did continue onward. Luna was a robotic craft series of missions that commenced in 1959 and employed both orbiters and landers. In 1959, Luna 1 missed impacting the Moon and ended up in a solar orbit. Also that year, Luna 2 impacted the Moon, becoming the first man-made object to reach the lunar surface. Just a month later, Luna 3 reached the lunar far side and returned the first images of the hemisphere of the Moon we can never normally see. Failures marked many of the other Luna missions.

One of Luna 9's 360-degree panoramas, showing nearby rocks and dust, and the horizon 1.4 kilometers (0.9 mile) away. The lander was initially at a 15-degree tilt, but the regolith (soil) on the Moon shifted underneath it and increased this to 22.5 degrees.

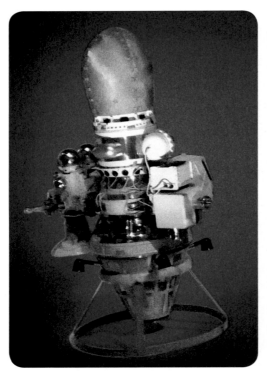 

Luna 9 performed the first-ever soft landing on the Moon on February 3, 1966. The lander was cleverly designed to right itself upon landing, standing upright no matter what rocky surface it landed on. Four protective "leaves" then unfolded, exposing the lander's panoramic camera and instruments.

But in 1966, the Soviets chalked up two great successes: Luna 9, in February 1966, achieved a soft landing on the Moon and was the first spacecraft to do so on any body other than Earth. Its five panoramas became the first close-up images of the Moon's surface. The following month, Luna 10 entered lunar orbit, becoming the first artificial satellite of the Moon.

In the United States, too, 1966 would be a key year in the increasing momentum of the Apollo program. In February the first test mission, known as AS-201, was launched. Using the IB launch vehicle, this unmanned flight provided the first validation of the entire production Block I Apollo Command/Service Module. It was a suborbital flight that lasted just over 37 minutes and demonstrated the service propulsion system and reaction control systems of each of the modules. It also successfully tested the Command Module's ability to return safely to Earth using its heat shield.

In July, NASA controllers conducted another test flight, designated AS-203, from Cape Kennedy. This was an unmanned flight of a Saturn IB booster without a Command

Seven radio sessions, totaling eight hours and five minutes, were transmitted as were three series of TV pictures.
The panoramic views were assembled from this data. Luna 9 survived for three days before its batteries ran out.

or Service Module. The purpose here was to test the design of the rocket's restart capability to boost astronauts from Earth orbit to a lunar approach. The test was successful, but after four orbits, some six hours into the mission, the stage was accidentally destroyed.

The second unmanned suborbital test flight of a Block I Apollo Command/Service Module launched with the Saturn IB rocket took place on August 25, 1966. Designated AS-202, the 1½-hour flight was the first one that included Apollo's Guidance and Navigation Control system and fuel cells. This test succeeded and American planners had a growing confidence that this system was ready to carry astronauts into orbit.

The first full-out test mission, designated Apollo 1, was planned for a February 21, 1967 launch, and a duration of 14 days. The first Apollo astronaut crew had been chosen and consisted of Gus Grissom, Ed White, and Roger Chaffee. Age 40, Grissom was an Air Force lieutenant colonel who was a veteran of the Mercury program. His successful mission, the second Project Mercury flight, was marred by the loss of *Liberty Bell 7*, which sank to the bottom of the ocean. He had also flown on Gemini III, when Alan Shepard was afflicted with an illness, thus becoming the first NASA astronaut to fly into space twice. The Apollo 1 mission would have made it three times for Grissom.

Liftoff! The Saturn 1B launched the Apollo spacecraft into Earth orbit in preparation for the manned flights to the Moon. After the completion of the Apollo program, it launched three missions to the Skylab Space Station in 1973 and was also used for the Apollo–Soyuz Test Project in 1975.

The Apollo 1 crew: Gus Grissom, Ed White, and Roger Chaffee pose at Cape Kennedy's Launch Complex 34 in January 1967, preparing for what should have been the first manned flight in the Apollo program.

Ed White was a 36-year-old Air Force officer, aeronautical engineer, and test pilot. As pilot of Gemini IV, he was a veteran of the program and had become the first American to walk in space, some three months after cosmonaut Alexei Leonov first achieved this feat. Roger Chaffee, age 31, was a naval aviator, aeronautical engineer, and a newcomer to spaceflight, having served as CAPsule COMmunicator (CAPCOM), chief of communications who speaks directly to the astronaut crew, for Gemini III.

The flight would employ a Saturn IB rocket and become the first manned test flight of the Apollo Command/Service Module in low-Earth orbit. Based at Cape Kennedy, the launch would be the first test of manned launch operations, ground tracking control facilities, and the Apollo-Saturn launch assembly. The Director of Flight Crew Operations, Mercury veteran Deke Slayton, selected the crew for Apollo 1. Originally, Donn Eisele was to be the third crewmember, but he dislocated his shoulder on a weightlessness training aircraft and was replaced by Chaffee. The mission's backup crew would consist of James McDivitt, Dave Scott, and Rusty Schweickart.

In the fall of 1966, NASA named Wally Schirra, Donn Eisele, and Walt Cunningham as the prime crew for the test flight that would follow Apollo 1. There were plans to

A Saturn V rocket leaves the huge Vehicle Assembly Building at Kennedy. The complete Saturn V stack with Apollo spacecraft on top was 111 meters (360 feet) high and 10 meters (33 feet) in diameter. To date, it remains the only vehicle to have carried humans beyond Earth orbit.

follow this subsequent test with an unmanned test flight of the LM (Lunar Module), also known as the lunar lander, and then a third manned mission would consist of two launched spacecraft that would test a rendezvous in orbit. The agency briefly considered launching the first Apollo test in conjunction with the final Gemini mission, but they scrapped this approach.

Apollo 1 would employ a capsule that was substantially larger and more complex than any previously used. The Command/Service Module had been designed by a group that was managed by Joseph Shea, a New Yorker who became an aerospace engineer and by 1963 was placed in charge of the Apollo Spacecraft Program Office. Shea had to approve aspects of the Apollo design and construction, and he had a difficult relationship with North American Aviation, the contractor building the capsule. At North American, Harrison Storms, a Chicagoan who became an aeronautical engineer, was in charge of managing the design of the Command/Service Module.

The critical development of the Apollo capsule, one of the most important aspects of the program, ran into difficulties throughout the early and mid-sixties. In NASA's corner, Shea, who famously kept a looseleaf notebook crammed with ideas and suggestions, thought North American had severe management difficulties. From North American's standpoint, Storms believed NASA delayed decisions too frequently and changed its mind on certain aspects of the program after decisions had been initially made. The result was an often-rocky working relationship between the space agency and its prime contractor.

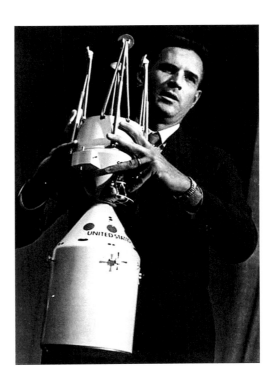

Head of the Apollo Program Office, Joseph Shea, demonstrates the docking of the Lunar Module and Command Module. He was deeply affected by the Apollo 1 fire in 1967 and left NASA shortly afterwards.

Chris Kraft at his console inside the Flight Control
area of the Mercury Control Center in 1966.

Moreover, Shea did not even get along well with all of the folks within NASA. Flight Director Chris Kraft described his unit's relationship with Shea as characterized by "intense animosity." While von Braun advocated a consensual working group, Shea, perhaps driven by insecurities, offered directives that often came off as dictatorial.

Nonetheless, Shea was described as a gifted engineer and NASA management put up with his catalog of peculiarities. Shea's reputation grew among those who were following the development of the Apollo program. Wernher von Braun, known as the key engineering genius behind the rockets, and Chris Kraft, the flight director, had already achieved something of a homespun, folksy, heroic reputation. Kraft, in fact, appeared on the cover of *Time* magazine, the newsweekly, in 1965. The publishers, Time-Life, famous for their coverage of the Mercury and Gemini programs, planned to place Shea on the magazine's cover in February 1967, the very month in which the first manned Apollo mission would fly.

Discussions about the Apollo capsule design raised concerns throughout the earliest phases of testing. The gaseous atmosphere of the capsule was designed to consist of pure oxygen. Oxygen is a highly reactive element and thus dangerous as a potential fire hazard. One of the earliest concerns the engineers discussed was the widespread use of Velcro in the crew cabin. The fastener, with its tiny, interlaced hook-and-loop structure, had been commercially produced in the late 1950s. Engineers worried about ripping off patches here and there and potentially creating a spark within the pure oxygen environment.

Shea himself later recounted saying: "Look, there's no way there's going to be a fire in that spacecraft unless there's a spark or the astronauts bring cigarettes aboard. We're not going to let them smoke." Shea simply told his engineers to "clean up" the spacecraft and make sure that the agency's fire safety rules were obeyed. As a joke, the three Apollo 1 astronauts — Grissom, White, and Chaffee — were sufficiently conscious of the safety concerns to present Shea with a photograph showing them posed over a model of the Apollo capsule, their heads bowed in prayer. "It isn't that we don't trust you, Joe," read the photo's inscription, "but this time we've decided to go over your head." With the hardware problems seemingly fixed to the satisfaction of NASA engineers, the craft underwent an altitude chamber test using Wally Schirra's backup crew on December 30, 1966. According to Jim Lovell, Schirra was not entirely pleased, and commented, "There's nothing wrong with this ship that I can point to, but it just makes me uncomfortable. Something about it just doesn't ring right."

Anticipation for the Apollo 1 mission was heavy, and the public looked forward to an exciting, new era of space exploration. The opportunity seemed to go hand in hand with the exploding social change of late 1966 and the dawn of 1967. From the point of view of the Americans, they seemed to be finding some momentum after nearly perpetually trailing the Soviets. A series of critical countdown tests was scheduled for January 27, 1967, at Cape Kennedy. The Apollo 1 spacecraft was rolled out onto launch pad 34. This was to be a "plugs out" test, to see whether the spacecraft would operate on simulated power while attached with all of its trimmings.

Should the plugs-out test work to the engineers' satisfaction, Apollo 1 would be declared ready for its February launch. Neither the capsule nor the rocket contained fuel, the explosive bolts had been disabled, and cryogenic (low-temperature) systems were absent. So the engineers felt there was little risk. At 1 p.m. EST, the three astronauts entered the Apollo 1 capsule with their pressure suits, were strapped into their seats, and Grissom noticed a "sour buttermilk" smell circulating through his air supply. The engineers could not find a cause, and the countdown resumed after a delay lasting a little more than an hour.

Three minutes after resuming the countdown, engineers affixed the hatch to the capsule door, which consisted of three parts — an inner hatch, a hinged outer hatch, and an outer hatch cover. The hatch cover was only partially bolted into place because of cabling running into the capsule to provide internal power. Hatches were sealed, and the environment inside was pressurized with pure oxygen.

From the control room, engineers could detect movements by the crew. Somehow, Grissom's microphone had been stuck open, and at one point he briefly remarked, "How are we going to the Moon if we can't talk between two or three buildings?" The countdown paused again. And then by 6:30 p.m. the clock remained at T-minus 10 minutes. The crew reviewed their checklist, and suddenly a momentary burst of voltage flashed into the capsule. After nine seconds, the open mic recorded in

The Apollo-era Mission Control Center, seen here in a View-Master stereo, was located in Building 30 at Johnson Space Center.

all likelihood Grissom screaming "Hey!" or "Fire!," followed by sounds of scuffling. Another astronaut, probably Chaffee, then said, "We've got a fire in the cockpit!" After nearly seven more seconds, a final transmission, garbled, seemed to say "We've got a bad fire — let's get out — we're burning up." The final words, perhaps coming from White, ended with a scream.

The intensity of the fire surged, fed by pure oxygen. The craft's inner wall ruptured. Intense heat and thick smoke billowed out of the capsule and prevented the ground crew's opportunity to rescue Grissom, White, and Chaffee. Some who rushed to try to save the crew feared the capsule might explode. After five minutes, the ground crew finally opened all three hatches. As smoke cleared, they found the badly charred remains of the astronauts. Grissom and White's suits and life-support hoses were melted. White, it appeared, had tried to open the hatch. It took nearly 90 minutes to remove the bodies. Autopsies revealed that all three had died of asphyxiation before their bodies were burned substantially by the fire. The fire likely started when an electrical arc occurred from a silver-coated copper wire that had been stripped of its insulation by repeated opening and closing of the spacecraft door.

The charred and disassembled remains of Apollo 1, weeks after the accident.

The Apollo 1 tragedy stunned the nation, shocked everyone within the space program, and set the Apollo program back substantially. It marked a low point in the American space program that would not be echoed with such tragedy until the *Challenger* and *Columbia* disasters, 19 and 36 years later, respectively. Hindsight, of course, makes determining failures relatively easy. But the decision to use a pure oxygen atmosphere for the Apollo capsule seemed, even as it developed, highly questionable. Even my own father, a chemical engineer who had worked on the Manhattan Project a generation before Apollo, would in later years simply shake his head and say that he couldn't imagine how NASA had approved running electrical wiring through a pure oxygen atmosphere. It was something that would have been unthinkable, he said, even back in the days of the Second World War. The Russians had had a similar tragedy in 1961 which resulted in the death of cosmonaut Valentin Bondarenko in an oxygen-rich atmosphere, but at that time they did not share even such potentially life-saving information with the Americans.

In the wake of the tragedy and an extensive investigation, NASA made changes. They modified the cabin atmosphere from oxygen to a nitrogen/oxygen mixture. They eliminated flammable cabin and spacesuit materials. They changed all manned flights to the Block II mode, using the actual lunar spacecraft with a quick-release hatch.

Now George Mueller, NASA Manned Space Flight Administrator, established a new set of goals for the upcoming test flights. By late spring 1967, NASA had its new plan in order. Reconfiguring the Command/Service Module in the wake of the Apollo 1 tragedy would allow the agency to catch up with its development and confidence in the engineering and testing of the lunar lander, the LM, and with the Saturn V booster.

In the summer of 1967, the world changed again. The evolution of music and counterculture exploded in the "Summer of Love." It was a tumultuous time, when things seemed to be reeling forward at an accelerating rate every week. More than 100,000 kids crowded into the Haight-Ashbury district of San Francisco, and rock 'n' roll, free love, and experimentation with drugs happened all over. Opposition to the Vietnam War grew by the week, and in the United States, occasional demonstrations cropped up on college campuses. The Monterey Pop Festival gave parts of the world their first glimpses of Janis Joplin, Jimi Hendrix, Jefferson Airplane, The Who, and countless other bands. The Beatles' *Sgt. Pepper's Lonely Hearts Club Band* provided a soundtrack for it all.

Following the Apollo 1 tragedy, a new and improved Command Module main hatch was designed and implemented so that it could be easily opened. The crews could now open the hatch in three seconds and egress in 30 seconds.

And through the summer, NASA pushed along, determined to get Apollo back on track. By November 9, 1967, the agency launched the next test flight, designated Apollo 4. It constituted the first unmanned test flight of the Saturn V rocket. It was a success, splashing down after nine hours.

No one knew it yet, but the year 1968 would be one of the most tumultuous in recent history. NASA commenced the year, January 22–23, with the next test flight, Apollo 5. This 11-hour experimental flight was also unmanned, and the first test of the Apollo LM, which would carry the astronauts down to the Moon's surface. By the spring of 1968, NASA conducted yet another test flight, Apollo 6. The nearly ten-hour flight, on April 4, demonstrated the trans-lunar injection capability of the Saturn V rocket, carrying a payload with about 80 percent the mass of the Apollo spacecraft. It also repeated a test of the Command Module's ability to survive a heat shield reentry. Despite reviewing some minor engine troubles, NASA's engineers were satisfied with the test and had high confidence in the program moving forward.

And, throughout it all, the year 1968 seemed to be boiling over. Early in the year, as the United States believed it was turning the tide in the Vietnam War, the surprise Tet Offensive sprang forward, with Viet Cong forces stunning and driving back US and South Vietnamese armies in chaos. Weeks later, US troops killed scores of civilians in the My Lai Massacre. In the United States, antiwar protests gathered even more momentum. Civil Rights clashes became increasingly violent. Befuddled by chaos, President Lyndon Johnson announced he would not seek reelection.

In April 1968, against a backdrop of anticipation for space exploration, the film *2001: A Space Odyssey* was released and mesmerized audiences. But the world seemed to be spinning out of control. The great civil rights leader Martin Luther King, Jr. was shot dead in Memphis, shocking the world. Weeks later, US presidential candidate Bobby Kennedy was shot and mortally wounded at the Ambassador Hotel in Los Angeles, dying the next day.

Toward the end of this uniquely chaotic year, NASA set its sights on the most important test flight yet. Apollo 7 would be the first Apollo mission to carry a crew into space. The mission launched on October 11 and lasted nearly 11 days. The crew consisted of Wally Schirra, Donn Eisele, and Walt Cunningham. The only manned flight to launch from Complex 34 at Kennedy, it would also be the last flight from that pad.

The flight was designed as an Earth-orbiting venture in which the crew would check out the redesigned Block II Command/Service Module with a live crew onboard. In this first American three-astronaut mission, Schirra was the overall Commander, Eisele was the Command Module Pilot, and Cunningham was the Lunar Module Pilot. The Saturn IB booster worked well, and mission operations worked as planned. The long duration, Schirra's worsening cold, and unhappiness with food and waste disposal made the crew somewhat cranky. But the objectives worked quite well, splashdown was flawless, and NASA was now ready, in another few months, for Apollo 8, which would orbit the Moon.

The end of 1968 brought more Earth-shattering changes in politics, war, social conscience, music, and culture. In the United States, Richard Nixon was elected President, and in England, the Beatles released their revolutionary *The Beatles*, known as the White Album, which would provide another underscore for the dizzy pace of the days to come.

The first manned Apollo flight, Apollo 7, took place in October 1968 and featured several tests of the Apollo spacecraft systems. One was to ensure that the Lunar Module adapter panels opened properly, and they are seen here "blooming."

The Apollo 7 crew, Walt Cunningham, Donn Eisele, and Wally Schirra,
left to right, pose for photographs on the launch pad.

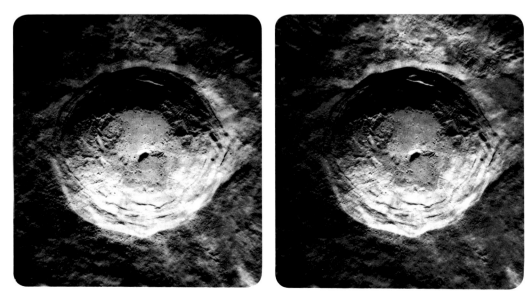

In the latter half of 1967, the American probe Lunar Orbiter 5 surveyed prospective Apollo landing sites. Among the areas imaged were
the prominent crater Aristarchus, a 40-kilometer (25 mile) diameter dish that sinks 3.6 kilometers (2.25 miles) into the lunar surface.

Valentina Tereskhova comforts Maria, daughter of cosmonaut Georgy Dobrovolsky, at the funeral for the three fallen cosmonauts of Soyuz 11 on July 2, 1971. The other crew members tragically lost were Viktor Patsayev and Vladislav Volkov.

5 THE SACRIFICE

By early 1969, it was clear to everyone on the planet that the world had changed. The innocence of the Eisenhower era, with its economic prosperity and blithe consumerism, was long gone. Social revolution had transformed the Western world and was creeping into other corners of the planet, too.

The Apollo program was now looking more likely to win the space race, with its ultimate goal of getting to the Moon. The Soviet program had stalled after the death of Sergei Korolev and other disheartening factors. And yet the Americans had also experienced their biggest setback yet, with the tragedy of the Apollo 1 fire. Several days after the tragedy, Gus Grissom, Ed White, and Roger Chaffee were buried at Arlington National Cemetery in Virginia. Months passed as a somber nation came to grips with the reality of spaceflight – an exceedingly dangerous business. In both the United States and the Soviet Union, engineers were pushing the boundaries, attempting things that had never been accomplished before – and that always involves risk.

Before we explore the successes of the momentous year 1969, let us pause to examine the large number of sacrifices that humans, and animals too, made in order to get our species into space, to lift us off of Mother Earth. The list is a long one, and one that was kept partially secret for many years. We've already encountered the terrible Nedelin catastrophe at the Baikonur Cosmodrome in Kazakhstan, in which at least 78 people died on October 24, 1960, when second stage engines of an ICBM ignited accidentally.

In the time between the Nedelin catastrophe and the Apollo 1 fire, other humans lost their lives in accidents. The first space-related fatality occurred on March 23, 1961, when Soviet fighter pilot Valentin Bondarenko, then a cosmonaut in training, participated in a test inside a low-pressure altitude chamber at the Institute of Biomedical Problems in Moscow. During the 10th day of a 15-day endurance experiment, with the chamber containing at least 50 percent oxygen, Bondarenko finished his testing for the day. He then cleaned his skin with an alcohol-soaked cotton ball, and tossed it aside. The cotton landed on an electric hot plate he was using to brew tea. The cotton ignited; Bondarenko tried to smother the flames with his sleeve, but that caught fire quickly in the oxygen-rich atmosphere. The pressurized chamber meant that a doctor, watching the tragedy in horror, couldn't open the door for about half an hour. Bondarenko's badly burned body was taken to a hospital, where the only veins available for an IV were in his foot. His fellow cosmonaut Yuri Gagarin watched his mortally wounded colleague at bedside, and 16 hours after the accident, the young pilot died. Less than three weeks later, Gagarin made his historic first spaceflight.

Three years later, an American test pilot lost his life. On October 31, 1964, Theodore Freeman, one of those selected in NASA's third astronaut group, was flying a T-38A Talon training jet from St. Louis, Missouri, to Houston, Texas, having attended a McDonnell training session in St. Louis. During his final approach toward the runway at Ellington Air Force Base in Houston on a foggy day, he ran into a cloud of geese in flight, and one flew into the port-side air intake of his engine causing it to flame out. Freeman attempted a runway landing but realized he was short and would crash into adjacent military housing. At the last minute, he banked away from the runway and pulled his ejection seat lever. The jet had rolled somewhat and he ejected in a near-horizontal attitude. He lacked sufficient time for parachute deployment and so he died as he impacted the ground, fracturing his skull and crushing his chest. Freeman became the first casualty of the NASA astronaut corps.

Earlier in 1964, a tragic accident had occurred at Cape Kennedy, unrelated to the Apollo program. Inside the spin test facility building, a Delta rocket was attached to the Orbiting Solar Observatory, an instrument designed to study the Sun from orbit. Eleven workers were in the room, and solid fuel in the rocket's third stage ignited prematurely. A spark of static electricity was later found to be the cause. All 11 workers were burned; eight survived, but three were critically burned and later died of their injuries. They were Sidney Dagle, Lot Gabel, and John Fassett.

In 1966, there was another American training jet crash, this time claiming two lives. Elliott See was an engineer, aviator, test pilot, and astronaut, having been chosen in the second group of NASA explorers. Charles Bassett was an electrical engineer and Air Force test pilot. He had been selected as one of NASA's third group of astronauts. In 1966, See and Bassett were both assigned to the Gemini program, and had been named as the prime crew for Gemini IX. On February 28, the two were flying for a training session to the McDonnell Aircraft facility in St. Louis from their normal home in Houston, and they were together in a T-38 Talon aircraft, See at the controls and Bassett in the rear seat. Tom Stafford and Gene Cernan, the backup crew for Gemini IX, were making the same trip in tandem with a similar aircraft, and it was a journey the astronauts had made many times. Weather at Lambert Field in St. Louis was bad, with scattered fog, snow, and occasional rain. The flight ceiling was 457 meters (1,500 feet), and broken clouds existed as low as 244 meters (800 feet). Both aircraft used an instrument approach, but when they cleared the approach they realized they had overshot the runway. See then circled around for a visual approach, but visibility was poor and deteriorating. Instead of following, Stafford pulled his aircraft up to attempt another instrument landing. See's aircraft circled around and dropped quickly, but left of the runway. He turned on his afterburner and pulled up, turning to the right, but it was too late — the T-38 struck the roof of the McDonnell building, cartwheeling and crashing into a parking lot. See was thrown clear and died on impact, while Bassett remained with the aircraft but was decapitated. Strangely and sadly, See and Bassett died within 152 meters (500 feet) of the spacecraft they would have piloted, as Gemini IX was at that time being assembled in a nearby building.

And then came a terrible tragedy in the Soviet space program. In April 1967, the Russians were ready for their first Soyuz flight, Soyuz 1, which was planned to be a complex and ambitious journey. The first crewed flight of a Soyuz, the flight would carry cosmonaut Vladimir Komarov into orbit, who would then rendezvous with Soyuz 2, to be piloted by Valery Bykovsky. The spacecraft would exchange cosmonauts before returning to Earth, but this complex plan was scrapped when the launch of Soyuz 2 was canceled due to thunderstorms.

It had been more than two years since the previous Soviet launch. On April 23, Soyuz 1 blasted off from the Baikonur Cosmodrome in Kazakhstan, and this first test of what would constitute a Russian lunar program moved forward. Problems occurred shortly after launch. A solar panel failed to deploy, depriving the spacecraft of power. Additional problems limited the maneuverability of the spacecraft. The Soyuz stabilization system was not functioning, and Komarov's manual override of the system was not entirely effective. After some 13 orbits, controllers decided to abort the flight. Following the 18th orbit, a day into the mission, Komarov fired the craft's retrorockets and entered Earth's atmosphere. He deployed a drogue parachute, followed by the main parachute. But the main chute did not unfold properly, and so Komarov deployed his last hope, a manually worked reserve chute, which promptly tangled with the drogue chute. Because of this, the Soyuz reentry module plummeted to Earth, striking the Orenburg Oblast in Russia, at a thunderous 140 kph (89 mph). Komarov was killed on impact.

When a rescue helicopter spotted the capsule on its side, the chute spread across the ground, the retrorockets suddenly fired, and by the time they reached the craft, it was on fire, dripping molten metal and with clouds of black smoke rising overhead. By the time the fire burned out, the capsule had completely disintegrated. They were able to recover Komarov's remains, which were taken to Moscow for analysis.

The tragic end of Soyuz 1, Komarov's death, the earlier loss of Korolev, and more trouble to come, pulled the Soviet lunar program back to a standstill. There were still plans on the slate, but the Soviets were losing momentum. Meanwhile, the Zond program (Russian for "probe") was aimed at information gathering on nearby planets and also scouting missions that would collect data on the Moon. In 1965, Zond 3 completed a circumlunar mission and became the second spacecraft to photograph the far side of the Moon. It passed by the Moon and after 33 hours lost communication with Earth, but it took 25 images of the lunar surface in visible light. With a minor modification, the Zond craft would be capable of carrying two cosmonauts. In 1968, the Soviets launched Zond 4, a robotic mission that was designed to test the worthiness of the craft for lunar missions. Through 1970, four more Zond missions were launched, all testing lunar capabilities, but they were only partially successful; and after 1968, Soviet leaders lost interest in the program. By 1970 it was canceled.

Initially, Soviet leaders also had plans for a crewed Moon landing and exploration program similar to that of Apollo. The Soviets planned to use what they called an N1/L3 rocket, a powerful counterpart to the Saturn V, which was developed beginning in 1965, four years after the commencement of the Apollo program. The project had been underfunded and was evidently rushed along after its late start; finally it was shelved after Korolev's death in 1966. Four attempts were made to launch an N1/L3, and they all failed. On the second try in 1969, the rocket fell back onto the launch pad and exploded, producing an enormously energetic blast.

The Soviet lunar plan called for a variant of the Soyuz craft, the Lunniy Orbitalny Korabi (LOK) command ship, to carry two men and three modules to make the lunar landing and exploration possible, in a heavier version of the Soyuz craft. The other component, the Lunniy Korabl (LK), would carry one cosmonaut, and so this would mean only one human would land on the Moon. The Soviets originally scheduled several launches to test this configuration beginning in 1967, but again, Korolev's death resulted in the program being substantially delayed.

Meanwhile, the Soyuz program, in the wake of the tragedy of Soyuz 1, was also not faring well. Before the launch of Soyuz 1, engineers had reported as many as 203 concerns over the design to party leaders, but these were ignored so the program could be pushed forward aggressively. The space race definitely played into this urgency, especially given the opportunity to gain ground on the Americans in the wake of the Apollo 1 disaster. The first human to experience space, Yuri Gagarin, had in fact been the backup pilot for Soyuz 1. His own concerns about the flight prompted him to try to get his friend Komarov pushed off the flight. Gagarin suggested that he himself be the pilot, believing that Soviet leadership would not risk his life, given his notoriety.

In October 1967, Alabama-born astronaut Clifton Williams died when his T-38 trainer crashed in Florida. In mid-November 1967, California astronaut Michael Adams died when his X-15 aircraft entered a violent downward spin during his descent toward Edwards Air Force Base, California. Little more than 10 minutes into the flight, the aircraft dove at Mach 3.9 and experienced 15 g vertically, breaking up the X-15's airframe.

Three weeks after the loss of Adams, yet another training jet crash cost the life of an American astronaut. Robert Lawrence, the first African-American astronaut, was a Chicagoan who became an astronaut in 1967 through the Air Force's Manned Orbital Laboratory program. On December 8 of that year he was flying in the backseat of an F-104 Starfighter, acting as the instructor for a trainee, Harvey Royer, who was flying the aircraft, learning how to make a steep-descent landing. The pilot came in poorly and the aircraft hit the ground, catching fire and rolling. Royer ejected and survived. Lawrence ejected sideways, killing him instantly.

A few months later, another tragedy occurred in the Soviet Union. On March 27, 1968, Yuri Gagarin and flight instructor Vladimir Seryogin took off in a MiG-15UTI, on a routine training flight from Chkalovsky Air Base northeast of Moscow. Near the town of Kirzhach, Vladimir Oblast, the jet crashed, killing both Gagarin and Seryogin. The cause of the crash is not exactly known and has been the subject of speculation for 50 years. The KGB investigated the crash, finding that airbase personnel provided Gagarin with outdated weather information, and conditions had deteriorated once he was airborne. At the base, ground crew members left external fuel tanks attached to the fighter, which Gagarin didn't plan for. The KGB concluded that a bird strike disabled the aircraft, or that Gagarin lost control because he had to make a sudden maneuver to avoid another aircraft. The outdated weather information led Gagarin to believe he was at a higher altitude than he actually was, and once the MiG started a spin, he lost control.

The loss of the first human being to reach space just seven years after his historic achievement struck the Soviet Union and the entire world very hard. Gagarin was a world hero, and his loss was a horrifying shock. His remains were cremated and interred within the wall at the Kremlin. The Soviet lunar plans, limping along as they were in the wake of the loss of Korolev and all the other setbacks, now turned toward Alexei Leonov, the world's first spacewalker, as the likely first Soviet to set foot on the Moon. But the program was now inching along and almost running out of gas. And with American successes to come in short order, the momentum for the Soviet Moon landing program would soon completely ebb away.

Over time, the toll of fallen astronauts and space workers has continued. In May 1968, William Estes, a pad worker at Cape Kennedy, was killed when a pressurized water line malfunctioned and struck him in the chest. In June 1971, the crew of Soyuz 11 were killed after undocking with space station Salyut 1 after their three-week stay. Georgy Dobrovolsky, Viktor Patsayev, and Vladislav Volkov died after a cabin vent valve accidentally opened during the separation of the service module with the station. The three cosmonauts tragically became the only fatalities to occur while in space.

Long after the era of Apollo, fatalities continued to happen. Of course the world was horrified when the Space Shuttle *Challenger* exploded 73 seconds into its flight on January 28, 1986, killing the entire crew: Gregory Jarvis, Christa McAuliffe, Ronald McNair, Ellison Onizuka, Judith Resnik, Michael Smith, and Dick Scobee. Seventeen years later the American shuttle program experienced its second devastating loss when *Columbia* disintegrated during reentry, killing the entire crew: Rick Husband, William McCool, Michael Anderson, David Brown, Kalpana Chawla, Laurel Clark, and Ilan Ramon. In 1993 cosmonaut Sergei Vozovikov drowned during water recovery training in the Black Sea. In 2014, American Michael Alsbury was killed when SpaceShipTwo VSS *Enterprise* crashed during a test flight.

Rocket explosions sometimes caused fatalities as well, as with the loss of nine lives in June 1973 and 48 lives in March 1980, both at the Plesetsk Cosmodrome in Mirny, Arkhangelsk Oblast, in the Soviet Union. In Xichang, China, in January 1995, a Long March rocket veered off course after launch, killing at least six people. The following year, at least six people were killed when a rocket veered away after launch and struck a village. The Chinese government reported six deaths; independent reports suggested at least 100 people died. At the Alcântara Launch Center Maranhão, Brazil, in August 2003, a rocket exploded, killing 21 people.

Let us not forget also the many non-human animals sacrificed on Man's journey to the Moon. None of them were given a choice, of course. This episode of history now looks shameful in its disregard for the dignity and well-being of hundreds of mice, dogs, and even monkeys, in the name of science. The most well-known tragic animal hero was Laika, the dog sent up on the Russian Sputnik 2 mission, in the certain knowledge that she would die in space. But the Americans too abused animals, and concealed the misery that their monkeys had to endure, in the pretence that these creatures enjoyed their notoriety.

Exploration, the gathering of knowledge, and pushing the boundaries of what it means to be a human, living on our planet Earth, has always been dangerous. To the astronaut-explorers, this risk and this cost has always been worth the reward of moving our species on its initial steps out into the cosmos. And the costs, the sacrifices, also had another effect — they bonded the astronauts and cosmonauts strongly together. Despite the fact that the space race was a phenomenon of the cold war, of hotly contested political rivalries, the space explorers themselves began to forge strong bonds in the sixties. Rivalries gave way to friendships. Competition gave way to cooperation. Anyone who ever saw Earth from space, whether it be from low-Earth orbit or from the Moon, would come to see our fragile blue planet as one world, with the political boundaries that divide nations appearing arbitrary and even ridiculous.

In one symbol of the shared humanity of space exploration, Apollo 15 astronauts carried a memorial to their fallen comrades to the Moon's surface. The Fallen Astronaut Memorial would come to symbolize a great deal about the lost explorers, and also about the shared goals of Americans, Russians, and many others. The full story of that remarkable artifact awaits you within the chapter on Apollo 15.

This bronze bust of Yuri Gagarin is the work of Russian sculptor Aleksei Dmitrievich Leonov (no relation to the cosmonaut), and was presented to the Science Museum in London by the International Charitable Fund "Dialogue of Cultures – United World."

Apollo 8 on the pad at twilight.

6 THE MOON CLOSE UP
(APOLLO 8)

By late 1968, NASA was ready for its most ambitious dress rehearsal mission yet. With the unmanned tests Apollo 4 and 6 behind them, and with the manned Earth-orbit flight Apollo 7 deemed a success, it was now time to step up the pressure. The original idea for what would be designated Apollo 8 was to test the LM in a low-Earth orbit at year's end. The original crew chosen for the mission included James McDivitt, Dave Scott, and Rusty Schweickart. At the same time the next crew in line, consisting of Frank Borman, Jim Lovell, and Bill Anders, was designated to fly Apollo 9, which would achieve a substantial test of the LM's flight capacity in a medium Earth orbit early in 1969.

But glitches in the development of the LM intervened. The complex equipment required for the assembly and testing of the LM meant that its development fell behind schedule. In the early summer of 1968, the LM intended to fly on Apollo 8 arrived at the Kennedy Space Center and was deemed inadequate, with several notable problems. The primary contractor assembling the LM, Grumman, had to re-engineer some aspects of the lander, and they adjusted their target dates to anticipate a delivery of springtime 1969. Thus, the deceased President's stated goal of landing on the Moon in the 1960s was now in jeopardy: NASA began to doubt whether a lunar landing could be achieved before 1970.

In midsummer 1968, the manager of the Apollo Spacecraft Program Office, George Low, proposed a potential solution. Low proposed flying a mission in December 1968 with the Command/Service Module configuration only, but instead of repeating the already-accomplished success of Apollo 7, this late 1968 flight could carry the Command/Service Module (CSM) all the way to the Moon, making this a circumlunar mission and potentially entering lunar orbit. This would allow the astronauts to conduct a dry-run of lunar landing procedures as they passed over the Moon's surface, accelerating those tests in the playbook from what was planned originally for Apollo 10.

NASA officials embraced the plan. It offered vital tests at the right time and would be dramatic in that it would carry a crew around the Moon and back home. The agency's chief, James Webb, needed some convincing. But after a short time Webb approved the mission change and Apollo 8 was established with a new set of goals. There would also be a new crew. Deke Slayton was still Director of Flight Crew Operations, and he determined to swap crews to enable Borman's to take over what would become Apollo 8.

The first crew to fly around the Moon, then, would consist of Commander Frank Borman, Command Module Pilot Jim Lovell, and Lunar Module Pilot Bill Anders. Indiana-born Borman, age 40, was a US Air Force test pilot who had flown on Gemini VII with Lovell.

Apollo 8 crew members Jim Lovell, Bill Anders, and Frank Borman pose beside the Apollo Mission Simulator at Kennedy Space Center, prior to the launch of the mission.

At 14 days, this was to set a new endurance record for time spent in space, and the craft also served as the target for Gemini VI-A, which marked the first rendezvous in orbit between two manned spacecraft. Meantime, Borman served on the official review board that examined the cause of the Apollo 1 fire. His testimony before Congress helped to convince that body to continue the Apollo program as a safe venture.

For the mission that became Apollo 8, Lovell was not originally chosen to be on the flight. But Michael Collins, the astronaut first chosen as the Command Module Pilot, needed back surgery after suffering a cervical disc herniation. So Collins missed out on Apollo 8, and in moved Lovell, in a reunion with his Gemini partner Borman.

The mission's Lunar Module Pilot, Bill Anders, was born in British Hong Kong in 1933, the son of a naval officer. At age 35, he was an Air Force fighter pilot, an electrical engineer, and a nuclear engineer. Selected in 1963 into the third NASA group of astronauts, Anders served on the backup crew of the Gemini XI mission and helped with NASA studies on the effects of radiation on space travel.

Now, Anders, along with Borman and Lovell, was poised to become one of the first three humans in history to travel to the Moon and back. In a NASA oddity, the mission's Commander, Borman, was less experienced than Lovell, and so Lovell became the first commander of a previous mission to fly as a non-commander.

Because of the new nature of this mission, special roles were assigned to the crew. Borman would act as Mission Commander. But Lovell, the Command Module Pilot, would act as navigator, and Anders, the Lunar Module Pilot, would act as engineer. On the ground, the communications specialists who would talk to the astronauts included Ken Mattingly and Vance Brand, who would fly later missions. Crew preparations for Apollo 8 began in simulators in September 1968, with Borman focusing on spacecraft reentry control, Lovell on navigation emergencies in case of a loss of communications, and Anders checking the state of the spacecraft's readiness.

The Soviet Union, meanwhile, had not abandoned its lunar-focused spacecraft activities. In 1968 the Russians pressed on with the Zond program, achieving their most unusual success yet. Zond 5 was launched on September 14, 1968, carrying the first living species to be propelled toward the Moon. Four days after its launch, the craft, which contained two tortoises, mealworms, flies, plants, seeds, and bacteria, circled the lunar surface at a distance of 1,950 kilometers (1,210 miles). This made the tortoises and the rest of the crew the first living creatures to pass around the Moon. On September 21, the capsule splashed down in the Indian Ocean and the creatures were recovered. The tortoises lost about ten percent of their body weight

The Saturn V F-1 rocket motors, each of which was capable of generating more than 1.5 million pounds of thrust.

during the flight but appeared to be active and had not lost appetite due to the experience. The Zond 5 flight helped to motivate NASA to push forward as soon as it could with the planned Apollo flights.

Preparations at the rocketry end also carried concerns. Apollo 8 would use the third model of the Saturn V booster, and the one designated for the mission was built in Kennedy's famous and enormous Vehicle Assembly Building in December 1967. Originally intended for an unmanned flight, this design proved problematic during the Apollo 6 test flight, when second stage engines failed and a third stage did not reignite. In the time leading up to Apollo 8, von Braun's team at the Marshall Space Flight Center conducted repeated experiments to alleviate the concerns. The engineers were concerned with so-called "pogo" oscillation, which is a violent vibration set up by instabilities in the burning of the liquid fuel. The engineers determined that the rockets vibrated at nearly the same frequency as the vibrations in the spacecraft, which caused something of a feedback loop that made the vibrations worse. They installed a system using helium gas to minimize the vibration problem.

The engineers were also concerned with the engine failure problems demonstrated in the Apollo 6 test. In a forensic analysis of the test flight, engineers found that a ruptured fuel line had caused a loss of pressure in engine number two. Problems compounded each other, too. When the second engine shut down due to an automatic shutoff, it also accidentally shut down engine three's liquid oxygen supply. This resulted from faulty wiring. The Marshall team re-engineered some parts, including fuel lines, igniter lines, fuel conduits, and other elements, and hoped this would eliminate the previous troubles.

By August 1968, the Marshall engineering teams tested their tweaks. They equipped a Saturn rocket with shock absorbing devices, which eliminated most of the pogo oscillation concern. They retrofitted a Saturn Stage II engine with fuel lines to demonstrate the necessary resistance to leaks and ruptures within the vacuum environment.

On September 21, 1968, the same day that Zond 5 splashed down, NASA engineers affixed the Apollo 8 capsule on top of their improved Saturn V rocket. On October 9 the Apollo 8 assembly made the slow journey, using a great "crawler" tractor at Kennedy, to Launch Pad 39A, a distance of 5 kilometers (3.1 miles). Launch for Apollo 8 was planned for December 21. NASA personnel continued to test the rocket and the craft right up until launch day. A final series of comprehensive tests took place on December 18, and the craft was deemed spaceflight ready.

Here we see (in stereo) the Saturn third-stage S-IVB on December 21, 1968, after its separation from the Apollo 8 spacecraft following trans-lunar injection. Attached to the S-IVB is the Lunar Module Test Article, which simulated the mass of a Lunar Module. Apollo 8 headed toward the Moon and the third stage entered a solar orbit.

Finally, more than seven years after John Kennedy's speech before Congress, NASA was ready to go to the Moon.

With worldwide anticipation centering on the mission, Borman, Lovell, and Anders climbed into the Apollo 8 capsule and readied for liftoff. The countdown proceeded well, and at 7:51 a.m., December 21, 1968, the Saturn V ignited, carrying the astronauts skyward. The three-stage rocket worked just as planned; the first and second stages peeled away, falling to the ocean below. The third stage boosted the craft into Earth orbit and stayed attached, as it would also provide the burn to put the capsule into a so-called trans-lunar injection, headed for a loop around the Moon.

The mission would last just short of a week. With the craft in Earth orbit, the crew onboard and crews on the ground spent 2½ hours checking to see that everything was in perfect working order. Following that checkout period, Michael Collins, acting as a CAPCOM on the ground, radioed the crew, "Apollo 8. You are go for TLI (Trans-Lunar Injection)." For the first time, humans received clearance to head for the Moon. After another 12 minutes, the third-stage engine ignited and performed the burn correctly, sending the crew out toward our nearest celestial neighbor.

Without the third stage, which had now been cast aside, the crew rotated the spacecraft to photograph the depleted third stage. As Apollo 8 inched its way toward the Moon, this configuration provided the first views of the entirety of Earth as seen by humans. Worried about the close proximity of the third stage to the Command/Service Module, Borman considered making a maneuver to separate the two but then decided against it. This consideration placed the mission about an hour behind schedule. The dilemma was solved when mission controllers concocted a plan to vent the third stage's remaining fuel, changing its path and placing it into a solar orbit, increasing its distance from the crew.

As time passed, the precision of the flight plan wavered still more. Seven hours into the flight, the crew found itself more than an 1½ hours behind schedule due to the concerns about the closeness of the third-stage booster. Additionally, Jim Lovell, as navigator, had to monitor the spacecraft's position so the crew could manually intervene with a return in case of a loss of communications. This meant that Lovell had to make star sightings with a sextant, and a gaseous cloud emanating from the nearby third stage obscured the view intermittently.

Moreover, the crew experienced for the first time what were then considered to be long-term spaceflight complications. Borman, Lovell, and Anders became the first humans to fly through the Van Allen radiation belts, and so they had to wear radiation dosimeters and carefully record levels of exposure. Further, the difference in temperature between sunlit and shadowed areas in the spacecraft was extreme. So the crew had to place the Command/Service Module into a slow roll that would expose all sides of the craft to direct sunlight more or less equally. The difference between sunlight and shadow could be as much as 300° C (500° F) — from oven baking temperatures to a hard freeze.

Prior to launch, NASA controllers mandated that at least one crewmember should be awake at all times. Some 11 hours into the flight, the crew had been awake for more than 16 hours, and Borman was first to make a shift asleep. But due to the radio chatter and spacecraft noise, he had trouble falling asleep. After a time, Borman secured permission from ground control to take a sleeping pill, but this helped only a little. Eventually he fell asleep but then awoke after a time and felt ill, becoming sick to his stomach. Borman's sickness spread small bits of vomit and feces into the cabin, and the crew attempted to clean it up. They made a record of Borman's condition and transmitted it to the controllers on the ground. In Houston, controllers held a conference with the crew and decided not to panic, as they believed Borman either had a reaction to the pill or a short-term illness like flu. In hindsight, historians now believe Borman was affected by a condition that makes about a third of space travelers ill during the first day as a reaction to weightlessness.

Aside from this drama, the cruise to the Moon was relatively uneventful. About 31 hours after launch, NASA held a televised event such that the astronauts could broadcast, in black and white, a greeting from space. The crew briefly delivered a tour of the spacecraft, which seemed cavernous compared with the earlier Mercury and Gemini capsules and described the view of Earth as they ventured farther away from it. The astronauts wanted to show Earth, but aiming the narrow-angle lens made that impracticable. Saturation of the incredibly bright Earth also made the best image they could obtain completely blown out, just a bright blob. After a 17-minute "show," Jim Lovell wished his mother a happy birthday to end the broadcast.

A striking view of Earth from the Apollo 8 spacecraft shows nearly all of the planet's western hemisphere and was made on December 21, 1968, most likely by Bill Anders. Visible are the mouth of the St. Lawrence River, including nearby Newfoundland, extending to Tierra del Fuego at the southern tip of South America. Nearly all of South America is covered by clouds.

Because of the spacecraft position and also its architecture, the crew could not see the Moon during most of the cruise phase. It didn't help that silicone sealant used on the capsule outgassed, creating a greasy, foggy oil that coated some of the spacecraft windows. The plan for sleeping in shifts also fell apart: Lovell and Anders finally got some sleep, but the careful planning of shifts proved impossible in the reality of spaceflight. Before they approached the Moon, the crew made a second TV broadcast, some 55 hours into the flight. They adapted the lens with filters so that the home planet could be seen by viewers without being blown out, and the crew described the view of Earth from such a long distance.

Nearly 60 hours into the flight, Apollo 8 left the gravitational sphere of Earth, and the Moon's gravity started tugging on the craft more forcefully. The crew continued working on navigation, calculating the trajectory that would carry them back home, and preparing to examine the Moon's surface when they had clear views of it. The crew readied for and conducted a burn, slowing the craft slightly and setting it up to pass just 115 kilometers (71 miles) from the lunar surface. And then came a critical moment, some 64 hours into the flight. The crew prepared for a lunar orbit

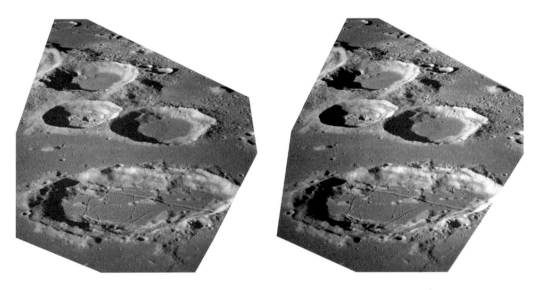

This oblique view of the lunar surface was taken from the Apollo 8 spacecraft looking southward toward the crater Goclenius and surrounding craters in the Sea of Fertility. Goclenius, the crater in the foreground with a rille-broken flat floor, is about 70 kilometers (44 miles) across. One rille crosses both crater rims and the central peak.

insertion, a maneuver that would take place on the far side of the Moon. At 68 hours into their flight, Mission Control informed the crew they were "a go" for their lunar passage and that they were "riding the best bird they could find." Jim Lovell said, "We'll see you on the other side," and for the first time, humans passed around the Moon and lost radio contact with Houston.

During their final check in preparation for the lunar orbital insertion, the crew caught their first glimpses of the Moon from the spacecraft windows. Shafts of sunlight cascaded downward and illuminated portions of the lunar surface, which seemed to spring alive with light, but in preparation for the maneuver, the crew couldn't yet focus on the sight.

Just after 69 hours into the mission, the burn took place and the Apollo 8 capsule slid into an orbit about the Moon. When Apollo 8 transmitted a signal again to Earth, it was Jim Lovell on the radio. He reported the first description of what the Moon looked like up close: "The Moon is essentially gray," he said, "no color; looks like Plaster of Paris or sort of a grayish beach sand. We can see quite a bit of detail. The Sea of Fertility doesn't stand out as well here as it does back on Earth. There's

A stereo detail of crater Keeler S on the lunar far side, from Hasselblad photographs taken by the Apollo 8 crew. This 30-kilometer (19-mile) wide satellite crater is west of Keeler itself, and is named after the American astronomer James Edward Keeler.

An oblique view of the crater Langrenus, on the eastern edge of Mare Fecunditatis, shows the steep terracing of the inner walls. "Quite a huge crater," said Jim Lovell when he spied it. "It's got a central cone to it. The walls of the crater are terraced, about six or seven terraces on the way down."

When these lunar far-side photographs were taken by the Apollo 8 crew, this remarkable crater was still unnamed, as were most of the far-side features. In 1979 it was named for French physicist Gaston Planté. The crater measures 37 kilometers (23 miles) across and is located on the eastern floor of the much larger Keeler crater, a portion of whose walls can be seen at lower right.

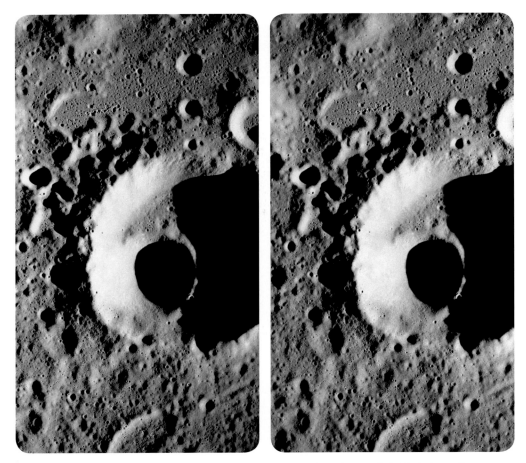

This peculiar far-side crater with a smaller crater on its floor is Korolev L, photographed with the Sun at a low angle by the Apollo 8 crew, Borman, Lovell, and Anders, who were the first humans to see the far side of the Moon with their own eyes. It stretches 31 kilometers (19 miles) across and lies in the Korolev Basin.

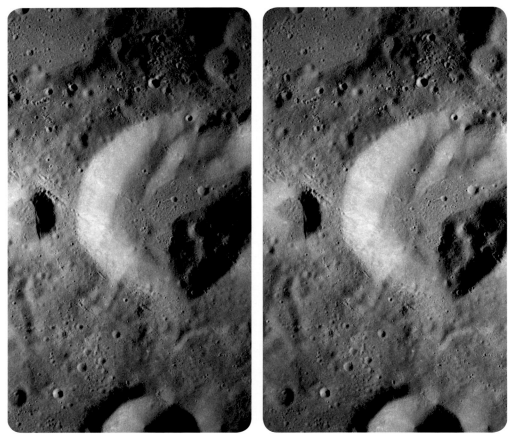

A rugged, heavily-cratered region between the craters Icarus and Amici. Photographed during the fourth lunar parking orbit, the same period during which Borman, Lovell, and Anders took pictures of the first-ever earthrise witnessed by human eyes, on December 24, 1968.

This iconic image, "Earthrise," was made by Bill Anders using a telephoto lens on December 24, 1968 and was the first such image taken by humans to show Earth rising over another celestial body. Earth floats some five degrees above the lunar horizon in the photo, with the spacecraft about 570 kilometers (365 miles) from the horizon. The white area on Earth, near the terminator, is the South Pole.

not as much contrast between that and the surrounding craters. The craters are all rounded off. There's quite a few of them, some of them newer. Many of them look like — especially the round ones — look like hits by meteorites or projectiles of some sort. Langrenus is quite a huge crater; it's got a central cone to it. The walls of the crater are terraced, about six or seven terraces on the way down."

As the craft passed over the lunar surface, Lovell and the others examined as much detail as they could, and they took photographs — 700 images of the Moon and 150 of Earth, the first time lunar features on the far side were photographed by humans. They examined potential landing sites like the one in the Sea of Tranquility that was chosen as the destination for Apollo 11. Borman focused on the spacecraft's condition and its ability to return to Earth. During the ensuing orbits, Borman read a small prayer for his church. The crew continued to check the status of the craft and observe all they could.

When the craft entered its fourth orbit, out from behind the Moon, the astronauts saw "Earthrise" — the full Earth rising over the lunar limb, for the first time ever. The resulting picture made at the time became an instant icon. Borman caught some sleep but awakened when it was clear that fatigue required all of them to work together. A few orbits later, the crew commenced a TV broadcast. During this, they made an impromptu reading of the Biblical Creation story from the Book of Genesis. Anders began by reading, "In the beginning, God created the heaven and the earth," and continued on, with Lovell and then Borman taking turns as well. It was a dramatic moment, and seemed to crystallize the sense of history happening during the flight. Borman closed with, "And from the crew of Apollo 8, we close with good night, good luck, a Merry Christmas and God bless all of you — all of you on the good Earth."

The spacecraft began its travel back to Earth on Christmas Day, December 25. It splashed down in the North Pacific Ocean, south of Hawaii, on December 27. For the first time, humans had traveled to the Moon and back. Humanity had reached out for the stars. All had gone well. NASA was now in full-on lunar mode. The Americans were nearly ready for the biggest show of all.

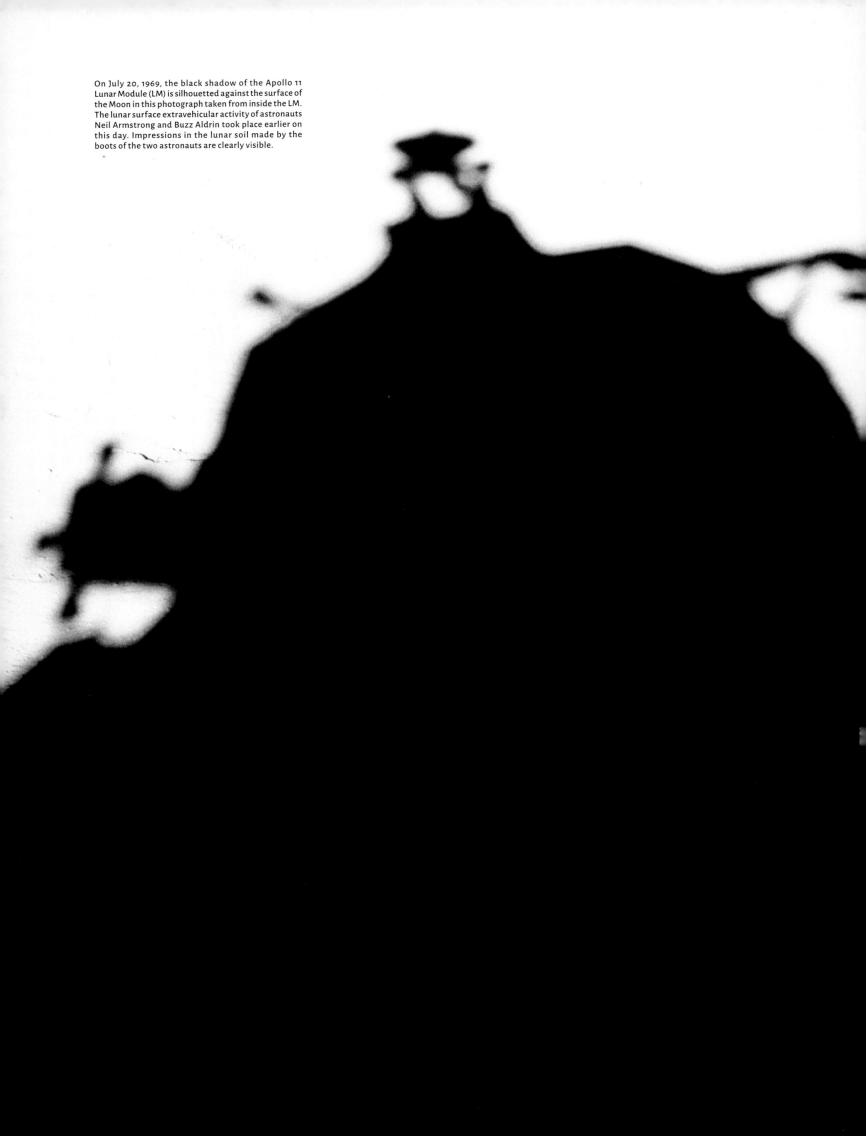

On July 20, 1969, the black shadow of the Apollo 11 Lunar Module (LM) is silhouetted against the surface of the Moon in this photograph taken from inside the LM. The lunar surface extravehicular activity of astronauts Neil Armstrong and Buzz Aldrin took place earlier on this day. Impressions in the lunar soil made by the boots of the two astronauts are clearly visible.

7 THE *EAGLE* HAS LANDED

(APOLLO 11)

The space race had now taken a decisive turn. Due to competitive agencies, the tragic deaths of several key figures, and political changes, the Soviet Union, which had led the early days of the race decisively, was now slowing down dramatically. Its lunar program had all but stalled. In the wake of Apollo 8, the American program was poised for the ultimate goal — John Kennedy's vision of sending a man to the Moon and returning him to Earth by decade's end. But first, two more test missions were needed to carefully prepare everything, to be absolutely sure all was a "go," for the lunar landing.

NASA planned its next test, dubbed Apollo 9, as a ten-day flight that would commence on March 3, 1969. This would be a check of the pairing of the Command/Service Module and the Lunar Module and would involve testing LM engines, backpack life-support systems, and spacecraft docking in low-Earth orbit. The Apollo 9 crew would consist of Commander James McDivitt, Command Module Pilot Dave Scott, and Lunar Module Pilot Rusty Schweickart.

This was the result of the crew swap with the original plan for Apollo 8, due to Deke Slayton's decision. McDivitt, age 39, was the Air Force test pilot who was a veteran of Gemini IV. Scott, age 36, was also an Air Force test pilot and had flown on the Gemini VIII mission. Schweickart, age 33, was an Air Force fighter pilot and aeronautical engineer who had been selected among the third group of NASA astronauts.

The crew swap had occurred in part because McDivitt wished to fly the LM mission, and Borman preferred the long circumlunar navigational flight. But this switch had a long shadow: it also meant that the crews to follow were juggled. By tradition, the backup crew for one flight would fly as the primary crew three missions hence. So this now placed the crew of Neil Armstrong, rather than Pete Conrad, in position to fly the first lunar landing mission, Apollo 11.

But first things first. On March 3, 1969, Apollo 9, the first complete test of the Apollo spacecraft assembly, blasted off from launch pad 39A at the Kennedy Space Center. The spacecraft achieved a low-Earth orbit ranging between 200 and 500 kilometers (125 and 300 miles) and commenced an array of tests and docking maneuvers.

The crew of Apollo 9, Jim McDivitt, Dave Scott, and Rusty Schweickart, pose at the Kennedy Space Center in front of the soon-to-be-launched Apollo 8 Saturn V rocket. Apollo 9 itself launched on March 3, 1968, and helped to pave the way for the Moon landing.

The Apollo 9 Command/Service Module photographed from the Lunar Module, *Spider*, during its Earth orbit, on the fifth day of the March 1969 mission. The docking mechanism is visible in the nose of the Command Module, *Gumdrop*, and floating in this "tin can" spacecraft was Command Module Pilot Dave Scott.

Apollo 9 tested the manned Lunar Module in Earth orbit as a self-sufficient spacecraft. This view, taken on March 3, 1969, shows Apollo 9's *Spider* Lunar Module still nestled in the third stage of the Saturn V rocket that carried it into space.

The crew conducted a docking, undocking, and re-docking of the Command Module and the Lunar Module, simulating what would be required to happen seamlessly in later missions in orbit about the Moon.

Both Scott and Schweickart performed extravehicular activities, spacewalks, Schweickart putting the Apollo spacesuit through its paces with its new life-support system, engineered to be independent of the spacecraft. While this occurred, Scott filmed him for posterity. Subsequently, Schweickart came down with a case of space sickness, which thwarted further planned activities outside the spacecraft.

Later on, Schweickart and McDivitt made a test of the LM, flying it and practicing maneuvers and docking with the Command Module. The two spacecraft reached a maximum separation of nearly 180 kilometers (more than 100 miles) before returning to close proximity and docking once again. Thus, the crew successfully tested all of the aspects of the mission they had planned and after ten days splashed down in the Atlantic Ocean, east of the Bahamas.

Now only one step remained for NASA before its conquest of the Moon — a full-fledged dress rehearsal. The mission designated Apollo 10 would launch on May 18, 1969, and last slightly more than eight days. This mission would aim for the Moon,

On March 7, 1969, the Apollo 9 Lunar Module, *Spider*, made its first autonomous test flight orbiting Earth. The crew onboard was the first ever to fly a spacecraft not capable of returning to Earth: re-docking with the Command/Service Module was essential for a safe return. These photographs were taken from the Command Module, *Gumdrop*. The landing gear on *Spider* has been deployed. Inside it are astronauts Jim McDivitt and Rusty Schweickart. Dave Scott remained at the controls of the Command Module.

The crew of Apollo 10, Lunar Module Pilot Gene Cernan, Commander Tom Stafford, and Command Module Pilot John Young, pose at the Kennedy Space Center.

as Apollo 8 had, and test all aspects of the lunar mission — including a separation of and descent of the LM — except for the landing itself. After Charles Schulz, the creator of the "Peanuts" comic strip, produced some artwork for NASA, the crew adopted "Peanuts" character names for the mission's craft: *Charlie Brown* for the Command Module and *Snoopy* for the LM.

During the fourth day of the Apollo 9 mission, astronaut Rusty Schweickart stands on the Lunar Module's deck during his spacewalk. This photograph was taken from within the Lunar Module, *Spider*, by Jim McDivitt.

Assembled from the first-ever photographs of a manned spacecraft over the surface of another world, this stereo view shows the Apollo 10 Command Module, *Charlie Brown*, as seen by astronauts Tom Stafford and Gene Cernan inside *Snoopy*, the Lunar Module, on May 22, 1969.

The Apollo 10 crew would comprise Commander Thomas Stafford, Command Module Pilot John Young, and Lunar Module Pilot Gene Cernan. For this important moment, the final test, all were chosen as spaceflight veterans. Stafford, age 38, was an Air Force test pilot and veteran of Gemini VI-A and Gemini IX-A. Young, also age 38, was a naval test pilot and veteran of Gemini III and Gemini X. Cernan, age 35, was a naval aviator and veteran of Gemini IX-A, flying with Stafford on that mission.

After they descended to 47,400 feet (14,500 meters) above the Moon's surface, Tom Stafford and Gene Cernan return the Lunar Module upward to dock with the Command Module during the Apollo 10 mission. This image was captured on May 22, 1969.

On May 18, the Saturn V launched Apollo 10 into space with a thunderous thrust, sending it into a trans-lunar injection maneuver and toward the Moon. Young rotated the craft to dock its nose to the top of the LM. For the first time, an Apollo mission carried a color TV camera in its payload, and the astronauts would make the first color transmission from space during the mission. After a cruise to the Moon that lasted three days, Young remained with the Command Module while Stafford and Cernan transferred to the LM and soon thereafter fired their descent engine. This carried the LM toward the lunar surface. They would approach an altitude of 15,400 meters (50,000 feet), the point at which the Apollo 11 astronauts would start a powered descent. They also visually surveyed the region where Apollo 11's LM would touch down, the Sea of Tranquility. Then they fired an ascent engine to return to the Command Module.

The Apollo 10 crew then encountered a problem. They had accidentally duplicated engine commands, and the LM began to roll as it headed upward. This happened during a live broadcast, and Stafford and Cernan panicked and shouted some nasty words as the craft spun at least eight times before they could recover control. The situation could have been grave, with the craft pitching toward the lunar surface.

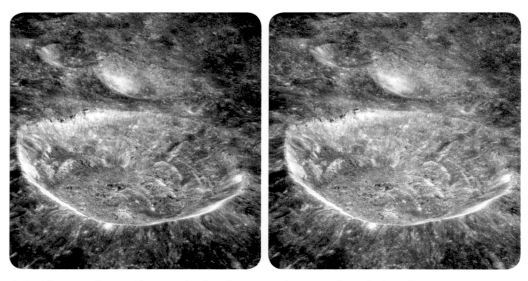

The far-side crater Necho, named for an Egyptian pharaoh, was captured in sequential stereo by the Apollo 10 astronauts in May 1969. This crater has an unusual asymmetric structure with a peculiar morphology and terraced walls on its western side.

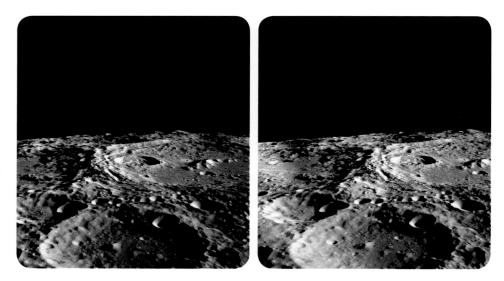

This hand-held view taken by the Apollo 10 astronauts shows the far-side craters Stratton (foreground), Keeler (top right), and Heaviside (top left). The rough terrain of the lunar far side is accentuated by the shadows caused by the low Sun elevation.

This tall Apollo 10 stereo view depicts the area between craters Lade and Rhaeticus and shows a chain of craters near Mare Imbrium. These vertically extended stereo pictures can be viewed best by sliding the OWL viewer up and down the page to scan the whole picture.

This photograph was taken by the Apollo 10 crew as a sequential stereo and portrays the crater Schmidt, 2.3 kilometers (1.4 miles) deep. Schmidt lies at the western edge of the Sea of Tranquility and stretches about 11 kilometers (6.9 miles) across, with bright walls and a very sharp rim showing few signs of wear.

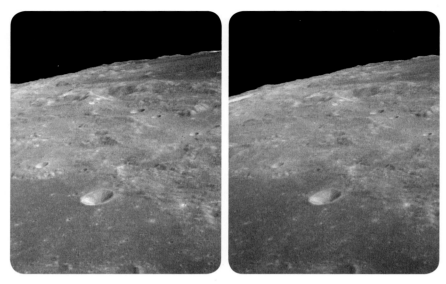

This highly oblique view taken by the Apollo 10 astronauts shows Mare Crisium and its adjoining highlands. Mare Crisium, the dark "sea" at upper right, is visible on the Moon's familiar near side with the naked eye, as a dark spot near the limb. This is the site where the Soviet Luna 15 robotic probe crashed on July 21, 1969, in an unsuccessful attempt to beat the US in returning a lunar soil sample to Earth.

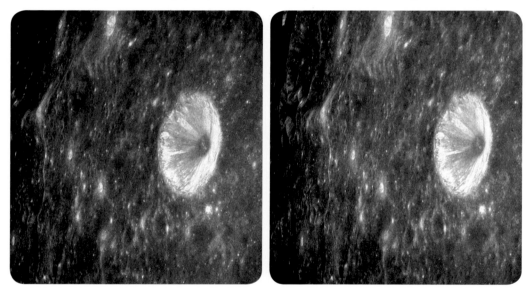

This Apollo 10 view shows an area at the eastern edge of Mare Smythii, a basin on the eastern limb of the Moon. A circular, bowl-shaped impact crater lies in the foreground, and it features a small floor and walls with a high albedo, making it appear very bright. A mountainous ridge is visible in the left background.

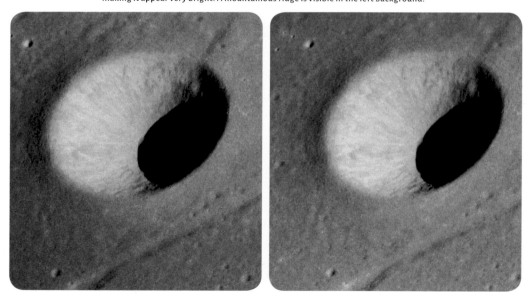

This oblique view of the crater Maskelyne G, a bowl-shaped crater in the southeastern part of Mare Tranquillitatis, was taken during the Apollo 10 mission. The six-kilometer-wide (3.75 mile) crater is just over 100 kilometers (62.5 miles) from what would become the Apollo 11 landing site. It was named for the fifth British Astronomer Royal, Nevil Maskelyne.

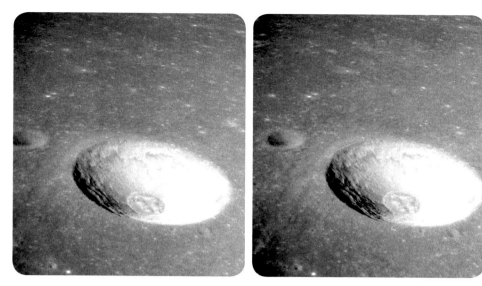

Messier B, satellite crater of the well-known oblong crater Messier in Mare Fecunditatis, shows well in this oblique close-up view made by the Apollo 10 crew. The crater's steep interior walls show their fine structure in the shadow. This stereo view also reveals that the crater has outer walls that slope gently into the surrounding mare.

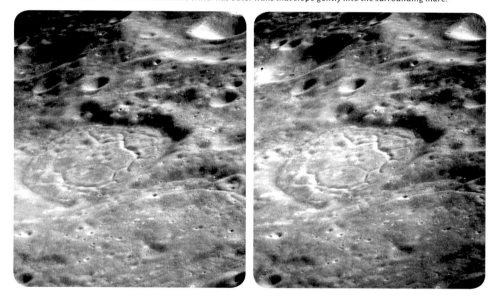

This oblique view was taken by the Apollo 10 crew while orbiting the far side of the Moon. It shows craters Tamm (foreground) and Van den Bos (center). Both are shallow with floors filled by solidified lava or impact melt from the Mendeleev basin some 225 kilometers (140 miles) to the northwest. The floor of Van den Bos is fractured with a series of rilles, created by cooling and hardening.

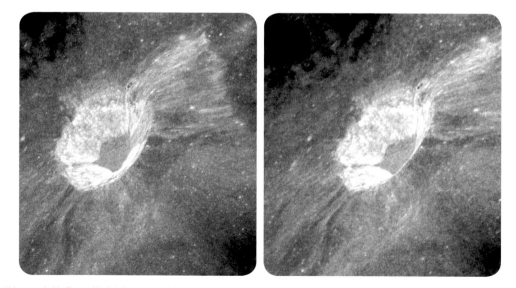

This remarkable flower-like bright crater with fresh rays of ejecta is Petit, named for French physicist Alexis Thérèse Petit, and photographed during the Apollo 10 mission. It lies on the northwestern rim of Mare Spumans (the Foaming Sea) on the lunar near side. The crater is five kilometers (3.1 miles) wide with a depth of one kilometer (0.6 mile). Craters with such bright rays are thought to be young.

The LM then fired further fuel and *Snoopy* headed toward a rendezvous with the Command Module. The reunion worked and all crew headed home, splashing down in the Pacific Ocean, east of American Samoa, on May 26, 1969. Despite the brief drama with the LM's ascent, the mission worked, all that was to be tested gained NASA's confidence, and the Americans were now fully ready for the big day to come.

The flight of Apollo 11, the mission chosen to attempt the first manned lunar landing, was scheduled for launch on July 16, 1969. The crew would consist of a second all-veteran, multi-purpose group, made up of Commander Neil Armstrong, Command Module Pilot Michael Collins, and Lunar Module Pilot Buzz Aldrin. For

The Apollo 11 crew, Commander Neil Armstrong, Command Module Pilot Michael Collins, and Lunar Module Pilot Buzz Aldrin, pose at the Kennedy Space Center prior to their historic mission.

In April 1969, Neil Armstrong participated in a suited simulation at the Manned Spacecraft Center in Houston, preparing for the upcoming Apollo 11 mission. In this image he wears a Hasselblad camera prototype mounted on his chest, and a lunar module mockup stands in the background.

each of the three, it would be his second time in space. Armstrong, aged 38, was an Ohio-born naval aviator and test pilot who had been chosen as one of the second NASA astronaut group. Along with Dave Scott, he flew aboard Gemini VIII in 1966. Michael Collins, also aged 38, was born in Rome, Italy, the son of a US Army major general, and became an Air Force test pilot before being selected to the third group of NASA astronauts. He was a veteran of Gemini X, along with John Young, flying in 1966. Aldrin, aged 39, was an engineer and Air Force pilot who was selected in the third astronaut group and flew, along with Jim Lovell, on Gemini XII, in 1966. Among the ground support crew for Apollo 11 were CAPCOMs Charlie Duke, Ronald Evans, Ken Mattingly, Bruce McCandless, Jack Schmitt, and Jack Swigert.

Mission planning went smoothly for the big show. Not everyone in NASA was amused by the informality of the names for the Apollo 10 Command Module and LM, however, and so more serious names were attached to the Apollo 11

In 1865 at the end of the American Civil War, Jules Verne wrote *From the Earth to the Moon*, one of the earliest science fiction novels and a satire of Americans. The projectile that was fired into space, shown here, was called the Columbiad.

counterparts — *Columbia* for the Command Module and *Eagle* for the LM. Although *Eagle* took its name from the feathered symbol of the United States, *Columbia* borrowed its name in part from the European historical name for the Americas, but also from the name for the ship from Jules Verne's 1865 novel, *From the Earth to the Moon*.

The anticipation for the Apollo 11 launch was immense. After years of planning, the Gemini missions, and all of the Apollo mission testing, the big day was finally approaching. In the summer of 1969, the sixties were rocketing toward a close, portraying a far different world than the Summer of Love in 1967 and the outright hippie culture of 1968. President Nixon ordered the first troop withdrawals from Vietnam. And following the death of President Dwight D. Eisenhower, he lay in state in the US Capitol. Students overran the administration building at Harvard University, protesting the war.

In pop culture, the Beatles gave their last public performance, on the London rooftop of Apple Records. The supergroup Led Zeppelin released its first album. Plans came together for the Woodstock Festival to be held in August in New York. Brian Jones of the Rolling Stones drowned in his swimming pool. Blind Faith played its first show in front of a massive crowd in London's Hyde Park. And the world's eyes increasingly turned toward that bright ball of light in our sky.

Launch day dawned clear and bright at Kennedy Space Center in Florida. It was Wednesday, July 16, 1969, and many

On July 16, 1969, the huge, 363-foot (112-meter) tall Saturn V rocket carrying the Apollo 11 crew launches from Kennedy's Launch Pad 39A at 9:32 a.m. EDT. Aboard the Apollo spacecraft were astronauts Neil Armstrong, Commander; Michael Collins, Command Module Pilot; and Buzz Aldrin, Lunar Module Pilot. Apollo 11 was the first lunar landing mission in history, and this view was made with a camera mounted on the launch tower.

thousands of people crowded along roads, around the public viewing areas, and even far off in other parts of the state to watch the towering flame of the rocket head skyward. The launch would be televised, and NASA's Jack King, Chief of Public Information, supplied commentary for the network coverage. With Armstrong, Aldrin, and Collins secured in the capsule, the mighty Saturn V rocket was slated for a launch at the now historic launch pad 39A, and the countdown began in the early morning. At 9:32 a.m. EDT, the rocket ignited and Apollo 11 lurched upward, with millions watching. The mission to land on the Moon was underway.

The launch appeared to be picture perfect, and within 12 minutes the craft entered Earth orbit, at an altitude of about 180 kilometers (100 miles). After an orbit and a half, the third-stage engine fired and moved the spacecraft into a trans-lunar injection, sending it toward our celestial neighbor. Some 30 minutes later the crew performed the maneuver that separated the Command/Service Module from the spent rocket stage, and allowed docking with the LM, extracting it, and discarding the rocket stage, sending the Command Module with the LM moonward.

The cruise toward the Moon lasted three days, and by July 19 the craft passed close enough to the Moon to fire its propulsion engine, setting it on a course orbiting the Moon. Armstrong, Aldrin, and Collins then orbited the Moon 30 times, and observed all manner of craters and other formations, paying particular attention to the region where they planned to land, the Sea of Tranquility. Previous robotic spacecraft had imaged the area and it appeared to be a safe landing site due to its flatness. The exact position would be about 19 kilometers (12 miles) southwest of the crater Sabine D, a 2.4-kilometer (1.5-mile) diameter, dish-shaped welt on the lunar surface that would later be renamed Collins in honor of the present crewmember.

On July 20, operations commenced that would lead to humanity's first steps on another world. Collins stayed aboard the Command Module, *Columbia*, as Armstrong and Aldrin climbed aboard the LM, *Eagle*, and began a descent operation. As the two craft separated, Collins carefully viewed *Eagle* as the lunar lander turned in front of him so that he could see any possible signs of physical damage to the craft.

Standing inside the cramped LM, Armstrong and Aldrin frantically checked readings and looked through the narrow window they had to peer out of. As they recognized landmarks, they found they were seeing them about four seconds ahead of the planned exercise, and they were going "long" – the spacecraft would pass

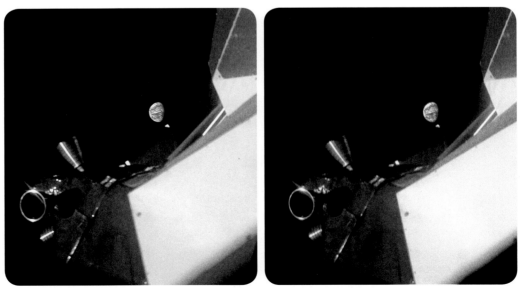

In this dramatic view made on July 19, 1969, before lunar orbit insertion, Earth lies beyond the Lunar Module, *Eagle*. The two images to make the stereo were shot from the Command Module, *Columbia*.

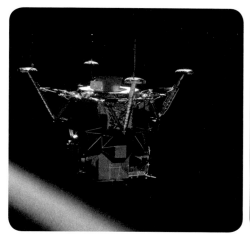

On July 20, 1969, the Lunar Module *Eagle* undocked from the Command Module and carried Neil Armstrong and Buzz Aldrin on their way to the lunar surface. This is one of a series of images made by Michael Collins from the Command Module as *Eagle* rotated, completing a visual inspection, looking for potential problems.

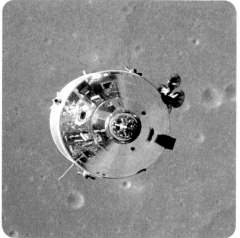

On July 20, 1969, the Command/Service Module, *Columbia,* hovered over the lunar surface, photographed from the Lunar Module, *Eagle.* The portion of the Moon visible below lies in the north-central region of the Sea of Fertility.

This view shows the perspective looking toward the docking tunnel of the Apollo 11 Command Module. Located at the craft's apex, it provided a passageway to and from the Lunar Module when the two components were docked. The 30-inch (0.76 meter) diameter docking hatch at the top of the tunnel acted as a pressure seal and a thermal barrier. It could be opened from both inside and out.

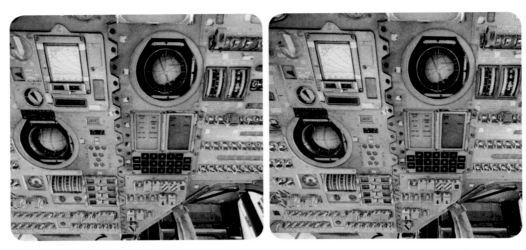

This is Neil Armstrong's view of the primary control panel in the Command Module, *Columbia*. The panel includes velocity, altitude and attitude indicators, as well as guidance and navigation controls. The spherical indicator at right is the Flight Director Attitude Indicator (also known as the "eight-ball"). It showed the spacecraft's attitude, rotation rate, and attitude error for all three vehicle axes.

the intended landing site. As they slowly moved downward, when they approached an altitude of about 6,000 feet, two program alarms sounded. The alarms resulted from the spacecraft computer recognizing that it could not accomplish all of the checks it should in real time. In Houston, at Mission Control, engineers were not overly concerned and provided the approval for the descent to continue. Later, NASA computer programming chief Margaret Hamilton wrote that the overload occurred due to a checklist error for the spacecraft preparation and that the computer automatically ignored lower-priority tasks in order to focus on the descent.

As *Eagle* continued to descend, Armstrong and Aldrin focused intensely, hoping to avoid any sort of mishap. Armstrong was consumed in piloting the craft, and Aldrin was calling out altitude marks to him. Back on Earth, Charlie Duke served as CAPCOM while he sat next to Jim Lovell and Fred Haise, and all were petrified in anticipation, worried over the landing. As the craft slowly lowered toward the

Space Capsule Communicators (CAPCOMs) at the Mission Control Center in Houston keep in close contact with the Apollo 11 crew during the historic mission and landing on July 20, 1969. They included, left to right, Charlie Duke, Jim Lovell, and Fred Haise.

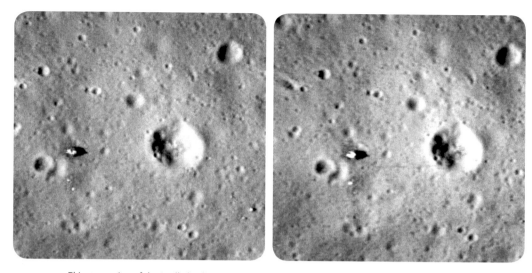

This stereo view of the Apollo landing site explains the difficulties Neil Armstrong faced as he piloted the LM down to what was expected to be a smooth landing plain. The image was made with the Lunar Reconnaissance Orbiter spacecraft on July 29, 2010, and shows the *Eagle* descent stage about 50 meters (162 feet) west of Little West crater, connected by a visible trail traced by Armstrong in his excursion.

lunar surface, moving downward with the speed of a hotel elevator, Armstrong peered outside and saw a boulder-strewn field, north of a crater later determined to be West Crater. He took semi-automatic control of *Eagle* and realized the craft had very little fuel left. The astronauts, in fact, received an erroneous "low fuel" warning due to the fluid sloshing in the tank and washing away from a sensor. But the fuel was running out. And those in Houston were very aware of the fact too. As the craft moved toward the surface — and no one knew exactly what the landing impact would be like — *Eagle* had less than half a minute of fuel remaining.

The spacecraft sank toward the Sea of Tranquility. A few moments before landing, Aldrin called out "Contact light!" as one of the hanging probes dangled onto the powdery lunar surface. Three seconds later, the craft landed squarely and Armstrong said, "Shutdown." After a few more words from Aldrin and Armstrong, Charlie Duke, in Houston, blurted out, "We copy you down, *Eagle*."

Moments later, Armstrong uttered, "Houston, Tranquility Base here, the *Eagle* has landed." Duke replied, with a bit of stumble, "Roger, Twan — Tranquility, we copy you on the ground. You got a bunch of guys about to turn blue. We're breathing again. Thanks a lot."

An iconic image! As part of his soil mechanics experiments, Buzz Aldrin took these pictures of his boot print in the powdery and soft lunar soil. The dry and fine-grained dust at Tranquility Base made the boot impression very clear as it compacted under the treaded sole. The image has often been mistaken for a picture of the legendary "small step" taken by Armstrong.

In one of history's most famous pictures, Buzz Aldrin stands on the lunar surface, gazing at the checklist taped to his wrist. In his helmet visor we see the reflection of Neil Armstrong taking the picture, as well as the *Eagle* and Buzz's shadow. Although many have assumed this spectacular photograph was carefully planned to capture an image of two astronauts and the LM, the moonwalkers modestly dismissed it as a lucky shot. Few have ever seen this famous picture in 3-D. The image was masterfully converted into 3-D by David Burder and re-edited here especially for this book.

It was July 20, 1969, at 4:18 p.m. EDT, and Charlie Duke wasn't the only one who was relieved. Many millions around the world were watching, and celebrations erupted when the news of the successful landing broke. Kennedy's vision had come true. Humans had landed on another world. And now the fun part was coming — the exploration. What would the Moon really be like?

About 2½ hours later, Armstrong and Aldrin prepared for their first walk on the Moon, as Collins orbited overhead in *Columbia*. They were supposed to sleep for

Armstrong and Aldrin had some difficulty erecting the first flag on the Moon, as the hard lunar soil proved to be difficult to penetrate with the flagstaff! The rod along the flag's top edge did not extend fully, causing the wavy flag to appear to be windblown. The flag planting was symbolic of the first landing on another world and was not intended to be a territorial claim by the United States.

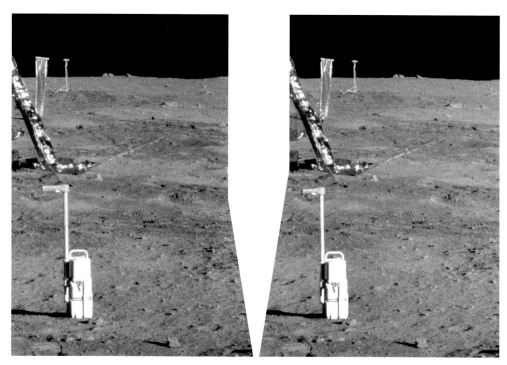

The Apollo Lunar Surface Close-up Camera (ALSCC) standing in front of *Eagle's* leg, on the lunar soil it is about to photograph in 3-D. It was the only stereo camera ever deployed on the lunar surface.

about five hours, but they did not, too excited to lose consciousness. Before readying for the moonwalk, Aldrin broadcast a message to those still on Earth: "This is the LM pilot. I'd like to take this opportunity to ask every person listening in, whoever and wherever they may be, to pause for a moment and contemplate the events of the past few hours and to give thanks in his or her own way." He then held a communion ceremony for himself, being an elder at his Presbyterian Church.

Peering out from *Eagle's* windows, Armstrong and Aldrin could see about a 60-degree wide field of the lunar surface, which appeared bright and gray. The pair intended to plant a US flag in the lunar soil and to deploy an experiment called the Early Apollo Scientific Experiment Package (EASEP), which would collect various data.

After more than two hours of preparation, at 10:39 p.m. EDT on July 20, Armstrong opened the LM's hatch and began to squeeze through it. At 10:51 p.m. EDT, Armstrong began his descent down the ladder to the Moon's surface, an event captured by the mounted slowscan TV camera on *Eagle*. At 10:56 p.m. EDT, Armstrong stepped onto the Moon's surface. Black-and-white TV pictures of the event were broadcast to countless millions of homes around the world; as a seven-year-old kid at the time, I can still remember the thrill of staying up "late" to watch this incredible moment.

As he slowly climbed down the nine rungs, Armstrong activated a switch to turn on the TV transmission and also uncovered an aluminum plaque with icons showing

Buzz Aldrin, Lunar Module Pilot, leaves the Lunar Module (LM) *Eagle* and descends the steps of the LM ladder as he prepares to walk on the Moon. This photograph was taken by astronaut Neil Armstrong, Commander, with a 70 mm lunar surface camera during the Apollo 11 extravehicular activity (EVA).

This Apollo 11 stereo view shows a rock measuring about 2½ inches (8 centimeters) across embedded in the powdery lunar soil. The small rocky fragments scattered around it suggest some erosion has taken place. Small pits, mostly smaller than an eighth of an inch, are visible on the rock, suggesting micrometeorite impacts. This image was made with the ALSCC, an instrument specifically designed to obtain the highest possible resolution of small areas.

Earth's hemispheres, the signatures of the astronauts, and a statement from Nixon: "Here men from the planet Earth first set foot upon the Moon, July 1969 A.D. We came in peace for all mankind."

About 6½ hours after the *Eagle* landed, Armstrong stepped into the powdery lunar surface dust, and proclaimed, "That's one small step for [a] man, one giant leap for mankind." The "a" wasn't audible in the transmission, altering Armstrong's quotation in many accounts, and Armstrong always maintained he said "for a man," and the "a" was probably obscured by static.

Using a sample bag, Armstrong then collected a lunar soil sample. Some 12 minutes later, Buzz Aldrin climbed down the ladder and joined his colleague on the Moon's surface.

The astronauts planted the US flag, described walking around in the light lunar gravity as "easy," and spoke to President Nixon via a telephone-radio hookup from the White House. The moonwalkers kicked up substantial amounts of lunar dust as they walked around and then focused on the experiments. They shot photographs,

Neil Armstrong beside the LM. Also visible are the US flag and the solar wind experiment which was conducted on all the Apollo missions.

Little West Crater, some 50 meters (162.5 feet) east of *Eagle*, is the spot Neil Armstrong visited on an unplanned excursion near the end of his moonwalk. Here he took pictures for a breathtaking panorama. The crater is about 30 meters (97.5 feet) across and four meters (13 feet) deep, and he took these images from the southwestern rim.

The foot of the *Eagle* lander standing on Tranquility Base lunar soil; stereo assembled from pictures taken by Neil Armstrong.

This narrow stereo view, looking southeast of the *Eagle*, was assembled from two sequential frames taken by Neil Armstrong for a panorama. Hollows and small craters dotting the plain are discernible in 3-D. The landing location in Mare Tranquillitatis was not as flat as mission planners had believed.

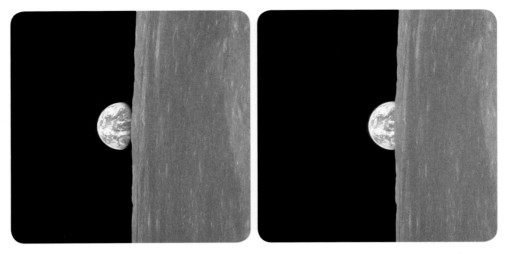

Earthrise as seen from behind the lunar limb in all its spellbinding beauty by the Apollo 11 crew before their legendary landing on the Moon. Our home planet provides an extraordinary spot of color beside the gray surface of Mare Smythii, on the lunar near side.

deployed a seismometer, set up a retroreflector, and collected rock samples. The lunar excursion was planned to last 34 minutes, but the tasks took longer than anticipated and so Houston granted the pair an additional 15 minutes. Armstrong strayed almost 60 meters (200 feet) from the LM during this first exploration of the Moon.

Aldrin entered the LM first, followed by his comrade. The two collected more than 21 kilograms (47 pounds) of rocks, and readied for their departure. After 21 hours on the Moon's surface, it was time to go home. Aldrin accidentally damaged a circuit breaker inside the LM, leading to apprehension about the engine. But after seven hours of rest, the pair awakened, fired the ascent engine, and lifted off from the Moon. They rendezvoused with Michael Collins and set a course back to Earth. The crew splashed down in the Pacific Ocean on July 24, returning as worldwide heroes.

The Moon had been conquered. John Kennedy's dream had been realized. A new era in human history had begun. And no one was exactly sure where this amazing moment would lead the world in its next stages.

This spectacular stereo of our home planet, taken on July 17, 1969, shows most of Africa and portions of Europe and Asia. It combines two views made from the Apollo 11 spacecraft when the ship was 98,000 nautical miles (181,500 kilometers) from Earth.

On July 24, 1969, the three Apollo 11 astronauts await pickup by a helicopter dispatched from the USS *Hornet*. The fourth man in the life raft is a US Navy underwater demolition team swimmer. All four men are wearing biological isolation garments. The crew splashed down about 812 nautical miles (1,503 kilometers) southwest of Hawaii and only 12 miles (19 kilometers) from the *Hornet*.

US President Richard M. Nixon was on location to welcome the Apollo 11 crew aboard the USS *Hornet*, prime recovery ship for the lunar landing mission. Astronauts Armstrong, Collins, and Aldrin are situated in the quarantine facility as the President looks on.

Released from their three weeks in quarantine, the Apollo 11 astronauts enjoyed ticker tape parades in New York (shown here), Chicago, and Los Angeles.

First Men: Neil Armstrong was painted by the legendary astronaut and Apollo 12 moonwalker Alan Bean. The painting depicts Neil Armstrong as he captured the iconic picture of Buzz Aldrin during the Apollo 11 mission. This version of the painting was made in 2007. Bean was celebrated for his use of color and texture. He very kindly granted us permission to use this painting shortly before he passed away in May 2018.

8 THE FIRST RETURN
(APOLLO 12)

Four months after the triumphant mission of Armstrong, Aldrin, and Collins, NASA was ready for its first follow-up act. Apollo 12 would begin where the previous mission left off, planning a longer and larger array of activities on the lunar surface. This now sixth manned flight of the Apollo program would concentrate on an area known as Oceanus Procellarum, the Ocean of Storms, a large mare near the western edge of the Moon's near side. This largest of all lunar maria covers a vast four million square kilometers (1.5 million square miles), and at its longest dimension stretches 2,500 kilometers (1,500 miles) from end to end.

The Apollo 12 crew drew on the large pool of available astronauts, and comprised Pete Conrad as Commander, Dick Gordon as Command Module Pilot, and Alan Bean as Lunar Module Pilot. This group brought a mixed array of experience to the new mission. Conrad, aged 39, was a native of Philadelphia, a naval officer and test pilot who had joined NASA's second astronaut group. He was a veteran of two Gemini missions, Gemini V and Gemini XI. The first astronaut from his group to fly a Gemini mission, he became the holder of a space endurance record, along with Gordon Cooper, when Gemini V spent just shy of eight days in orbit. On the Gemini XI mission, Conrad, along with Dick Gordon, docked with an Agena Target Vehicle soon after they reached low-Earth orbit. This mission was notable for testing the maneuver, which simulated the kind of docking that would be required for the Apollo Command Module and LM, and it was significant for achieving the highest Earth orbit ever — with an apogee, or most distant point in its orbit, of 1,369 kilometers (851 miles).

The crew of Apollo 12 prepares to make the first return trip to the Moon, following the voyage of Apollo 11. They pose here at the Kennedy Space Center, with Commander Pete Conrad in front, Lunar Module Pilot Alan Bean at the back, and Command Module Pilot Dick Gordon to the right.

Dick Gordon was also a spaceflight veteran. Forty years old, Seattle-born Gordon was a naval officer and aviator, test pilot, and chemist. One of NASA's third astronaut group, Gordon cut his teeth as backup pilot for the Gemini VIII mission before being assigned to Gemini XI, along with Conrad. Gordon and Conrad were selected to fly on the Gemini mission together, and also on Apollo 12, in part because they were close friends. They had even been roommates on the aircraft carrier USS *Ranger* during their naval careers in the 1950s. While the docking maneuver with the Agena vehicle in the Gemini XI mission was underway, Gordon conducted two spacewalks, during which he secured himself by attaching a tether to the Agena vehicle.

Apollo 12 marked the first flight into space for the LM Commander, Alan Bean. Age 37, born in Wheeler, Texas, Bean was a naval aviator, aeronautical engineer, and test pilot. A fellow member of Gordon's from the NASA astronaut group 3, he was selected as Backup Commander for Gemini X and then grew somewhat frustrated by his exclusion from the early Apollo flights. He became highly dedicated to astronaut training procedures, and when his classmate Clifton Williams was killed in an air crash, Bean replaced him as a backup crewmember on Apollo 9. By protocol, Williams had also been targeted to fly on Apollo 12. As Mission Commander, Conrad, who had known Bean earlier and liked him, asked NASA to replace Williams with Bean, which they did. In later life, long after Apollo 12, Bean was highly celebrated by fans of the space program for his wonderful paintings centered on Apollo 12 and other missions. Describing how his astronaut career produced scenes that few had seen, he possessed a special drive to share this vision with others. His paintings are now regarded as treasures associated with NASA and the space program.

Apollo 12's backup crew would consist of Dave Scott, Al Worden, and Jim Irwin, who were to fly together to the Moon as the primary crew for Apollo 15.

The launch of Apollo 12 was scheduled for November 14, 1969, a date pushing close to the end of the sixties, but the deadline for reaching the Moon set by John Kennedy's original motivating speech had already been fulfilled. By now, Western culture had swung completely upside-down, with the United States looking for a way out of the Vietnam War, and hippie optimism fading away in the wake of the summer murders in Los Angeles by the Manson Family. The last gasp of hyper-rock festivals was played out in the summer of the Woodstock Festival in New York, which drew half a million kids into the country for three days of peace, love, and music.

US President Richard Nixon looks on toward the launch of Apollo 12 at the Kennedy Space Center, as NASA Administrator Thomas Paine shields First Lady Pat Nixon with an umbrella.

But the spirit of NASA's quest for the Moon continued on apace. Launch day was not optimal, with a rainstorm hovering over the Kennedy Space Center, but that did not thwart the crew or the support staff. Conrad, Gordon, and Bean loaded into the Command Module and readied themselves for an on-time liftoff. The President of the United States, Richard Nixon, was on hand in Florida to watch the launch for himself. The call signs for the spacecraft components were traditionally named, with the Command/Service Module designated *Yankee Clipper* and the LM described as *Intrepid*. At liftoff, the mighty Saturn V lurched skyward and, just 36 seconds into the flight, a bolt of lightning shot through the rocket into the plume of ionized gas trailing below and on down to the ground. The rocket itself triggered the lightning discharge.

The lightning bolt created a problem. Inside the Service Module, three sensors detected the charge as an overload, which it actually wasn't. This detection, however, automatically knocked three fuel cells offline and also shut down much of the instrumentation inside the Command Module. And the trouble wasn't over. Just 16 seconds later, the rising rocket was hit by another lightning strike, which knocked out the so-called "eight-ball" attitude indicator inside the Command Module. Communications also became a problem. The stream of telemetry heading to Mission Control in Houston carrying important data back and forth to the spacecraft found itself in an unintelligible garble.

Despite this unfortunate pair of electrical events, the spacecraft continued on the proper course. The lightning strikes had not adversely altered the Saturn V Instrument Unit, which kept operating properly and kept the craft on target. Inside the Command Module, however, the crew were not amused. The loss of the fuel cells meant the Command Module was running on battery power, and this changed the power supply to the craft and activated nearly every alarm and warning light on the spacecraft's control panel. On the ground, a crew support member, John Aaron, believed he recognized the pattern of malfunctions from an earlier test and he suggested "Try SCE to aux," which meant switching the spacecraft's Signal Conditioning Electronics unit to auxiliary power. None of the others working on the problem — Conrad, Flight Director Gerry Griffin, CAPCOM Gerald Carr, or others — had recognized this potential solution. However, Alan Bean, the one crewmember on his first flight, also remembered this from a simulated failure in a training mission.

Bean switched the fuel cells back online and the spacecraft began to right itself in terms of checks and alarms. This quick

The Saturn V rocket carrying the Apollo 12 astronauts lunges upward into heavy clouds and intermittent thunderstorms on November 14, 1969.

action, taken by Bean and signaled by Aaron, saved the mission from a potential abort just a short time into the flight. The telemetry came back, communications were restored, alarms silenced, and data systems were back in shape. The mission quickly restored itself and the confidence of the crew and ground support flooded back across the spectrum. The spacecraft entered Earth orbit and then the astronauts slowly and carefully checked the spacecraft systems before they initiated a trans-lunar injection, firing the Saturn's third-stage rockets to carry the vehicle into a trajectory toward the Moon.

The careful spacecraft checks assuaged some serious fears. Other doubts, however, hung out there still. Controllers worried that the lightning strikes could have compromised the parachute package, firing it early, and thereby leaving no way for the returning astronauts to get home safely. In this scenario, the Command Module would plummet into the ocean in an uncontrolled, high-velocity dive, killing the crew. Controllers felt this was an unlikely possibility, and they did not tell the crew about it.

A malfunction also occurred with respect to the Saturn third stage, which was normally sent into a solar orbit after the separation of the Command Module and the LM, pulled into this orbit by the Moon's gravitational field. The third stage flew past the Moon at too high an altitude because of an error in the guidance system. After passing the Moon, it stayed in a semi-stable Earth orbit, from which it escaped only two years after the mission. It fell back into an Earth orbit in 2002, was spotted by an amateur astronomer who believed it to be an asteroid, and was finally identified as an artifact of the Apollo 12 mission, some 33 years after the fact.

Following their four-day cruise to the Moon, the crew of Apollo 12 approached the lunar surface. While Gordon remained in orbit inside the Command Module, Conrad and Bean entered the LM, which separated from the Command Module, and descended toward the target area in Oceanus Procellarum. The approach to Apollo 12's landing was in stark contrast to that of Apollo 11. In the former mission, Neil Armstrong had had to use manual controls to fly the LM down along an extended path to avoid an unforeseen boulder-strewn area and barely made it to a safe landing point before depleting the available fuel. By contrast, Apollo 12's descent was a feat of precision targeting. The area was one of prior interest, as well.

This region of the Ocean of Storms had been visited by several unmanned spacecraft, including the American missions Surveyor 3 and Ranger 7, as well as the Soviet Luna 5. The site would come to be called Statio Cognitum on maps of

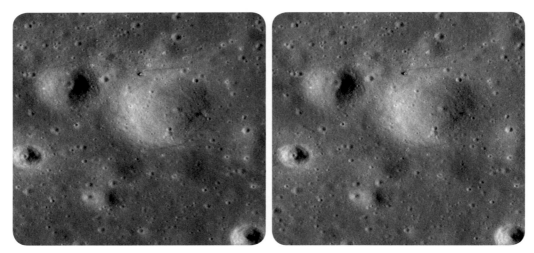

This image of the Apollo 12 landing site was captured by the Lunar Reconnaissance Orbiter on November 19, 2010. It shows the descent stage of the LM *Intrepid* next to the rim of the crater Surveyor. Inside the crater, on the right side, is the *Surveyor* 3 probe, which casts a shadow, and sunlight glints off the Apollo Lunar Surface Experiments Package northwest of *Intrepid*.

On November 19, 1969, the Apollo 12 Lunar Module *Intrepid* appears floating over the lunar limb as photographed by Dick Gordon from inside the Command Module, *Yankee Clipper*. The large crater occupying most of the lower half of this image is Ptolemaeus (with the smaller crater Ammonius within it), and second largest is Herschel at the right-hand edge of the photograph. At this time the LM was 110 kilometers (69 miles) above the Moon's surface.

the Moon, following the International Astronomical Union's naming of this small region as Mare Cognitum, meaning the Known Sea.

Conrad, true to his good sense of humor, named the region of the intended landing "Pete's Parking Lot." The LM's descent was automated, with Conrad taking over for manual adjustments near the final portion of the landing. As with Apollo 11, the LM descended slowly toward the Moon's surface, roughly at the rate of a hotel elevator. The intended landing site was chosen to be near the Surveyor 3 probe but far enough away to ensure that the lunar regolith blasted upward by the descent engine would not damage the earlier probe. Conrad, however, became somewhat concerned with the "roughness" of the surface during final descent and so adjusted the landing site to be short of the intended landing area — the "Parking Lot."

In the end, on November 19, 1969, *Intrepid* landed 183 meters (600 feet) from Surveyor 3, and the touchdown *did* "sandblast" the older spacecraft, which had robotically landed on the Moon two years earlier. Surveyor 3 was one of several US robotic craft to land on the Moon during the developmental phase of the Apollo program. It came down in a small crater and, because its engines failed to cut off at the elevation of

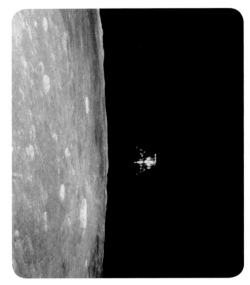

The Apollo 12 Lunar Module *Intrepid* looks like a mosquito flying above the Moon's surface on November 19, 1969. On board, astronauts Pete Conrad and Alan Bean were on their way to landing in the Ocean of Storms. These images were captured by Dick Gordon, from the Command Module, shortly after undocking.

An American flag, erected by the Apollo 12 crew, stands next to a small crater near the LM. A quick comparison with the flag planted by Apollo 11 astronauts shows that something went wrong here: the latch mechanism made to hold the crossbar broke, allowing the flag to droop. For later missions, NASA adjusted the supporting hardware.

four meters (14 feet), the craft bounced twice on the Moon, rising up again some 11 meters (35 feet), before coming to a settled position. The craft carried a surface soil sampling scoop, with which it dug several small trenches, scooped up soil, and using its TV camera mounted on board sent close-up pictures of the lunar soil back to Earth. The mission's scientific experiments lasted for two days and 17 hours.

Apollo 12 landed so close to Surveyor 3 that the crew could walk over to the older spacecraft. It thus became the first mission to visit, with human explorers, an older probe. Strangely, the sandblasting with lunar regolith actually cleaned the surface of Surveyor 3. The craft had a tan hue, owing to a layer of lunar dust, but the sandblasting exposed pure white areas showing the spacecraft's original finish.

After their November 19 touchdown, Conrad and Bean rested and prepared for the mission's assignments. In contrast to the short stay of Apollo 11, the pair would spend about a day and a half on the lunar surface, conducting two extravehicular activities that amounted to a total of just over seven hours and 45 minutes spent outside the LM.

The first moonwalk commenced about five hours after the LM touched down and lasted just four minutes under four hours. Conrad walked down the LM's ladder first and thus

Commander Pete Conrad, photographed by Alan Bean, descending the LM *Intrepid* ladder, about to become the third man to walk on the Moon, on November 19, 1969. He was about to say the legendary words, "Whoopee! Man, that may have been a small one for Neil, but that's a long one for me!"

Two spacecraft on the Moon: Surveyor 3 in the foreground, which soft-landed on the Moon in April 1967, and the LM *Intrepid* in the background. In November 1969 *Intrepid* landed some 200 meters (650 feet) from Surveyor 3, in the Ocean of Storms. In this image Pete Conrad is examining the Surveyor probe before detaching pieces of it to bring back to Earth for study.

Astronaut Alan Bean holds a special sample container filled with lunar soil collected during his moonwalk. This excellent picture was taken by Pete Conrad, who is reflected in Bean's helmet visor. Also, note the lunar dust on Bean's Hasselblad camera!

became the third human to walk on the Moon. Again he demonstrated his sense of humor, paying homage to Neil Armstrong, blurting out, "Whoopee! Man, that may have been a small one for Neil, but that's a long one for me!" The joke in part referenced Conrad's somewhat short stature, and had been orchestrated: the third man to walk on the Moon made a $500 bet with a reporter that he would say that line after she had asked him about Armstrong's famous declaration.

Bean followed, and the two accomplished a great deal scientifically during the two moonwalks. The second one took place 11 hours after the end of the first and lasted about three hours and 50 minutes. Conrad and Bean indeed walked over to the Surveyor 3 craft and disassembled pieces of the spacecraft, carrying off ten kilograms (22 pounds) of material to take back to Earth, including the probe's camera. When the camera was analyzed back on Earth, reports circulated that it contained live bacteria called *Streptococcus mitis*. This common bacterium is widespread on Earth, and the finding puzzled NASA researchers. At first the agency claimed that the camera had not been properly sterilized before being sent on the Surveyor 3 mission, maintaining that the bacteria were present on the Moon. However, they later retracted this claim, changing their explanation to contamination by workers

Alan Bean operating his Hasselblad camera on the Moon's surface. He was aiming his camera at the LM *Intrepid* when Pete Conrad took these images, made as part of a panorama.

Pete Conrad is shown here operating the Lunar Equipment Conveyor (LEC) outside the LM *Intrepid*. Because the idea for it came from clotheslines on pulleys outside New York City apartments, the LEC was nicknamed the "Brooklyn clothesline" by the astronauts. It was a simple device that made pulling down equipment from the LM and moving sample boxes up into the LM far easier. Conrad also wears his Hasselblad camera in this shot.

who examined the camera after its return to Earth. They pointed out that the camera came back in a porous bag, workers had their bare arms exposed as they examined the camera, and other sloppy procedures were employed. The bottom line is that the camera probably *was* exposed to bacteria after its return to Earth.

The science experiments performed by Conrad and Bean were exciting and set the stage for further and more complex experiments to come with the remaining Apollo missions. They set up the first ALSEP (Apollo Lunar Surface Experiments Package) experiment, a package of sensors that would relay long-term data about the Moon back to Earth. The astronauts measured the lunar magnetic field, made measurements of seismic data, and measured the strength of the solar wind. Conrad and Bean also carried the first color TV camera onto the Moon, attempting to increase the quality of television transmission back to Earth substantially. But when Bean carted the camera to a position near the LM and began to set it up, he accidentally pointed the camera toward the Sun, which destroyed the camera's primary tube, thus ending the TV coverage of the mission right away.

As with all lunar missions, the astronauts collected substantial numbers of Moon rocks to carry back to Earth. Analysis of the first returned sample indicated

Pete Conrad stands on the south rim of Surveyor Crater in this view, in the Ocean of Storms, sampling rocks from a small impact event with a sampling scoop. On his left is a carrier with more geological tools and sample bags, and a gnomon is on the ground near his feet. The image of the photographer, Alan Bean, is reflected in Pete's visor.

that Moon rocks were far more similar to Earth rocks than had been previously believed. This would remain a mystery for a time, and ultimately led to a stunning hypothesis about the origin of the Moon, namely that the Earth was struck by an unnamed planet and the resulting debris, including some of the Earth's, formed a disc around the Earth which eventually coalesced to form the Moon (see Chapter 14). The astronauts also took large numbers of photographs, although Bean accidentally left some behind when he departed, so exposed rolls of Apollo 12 film sit on the Moon's surface to this day.

After their lunar activities Conrad and Bean rested in the LM and then, on November 20, ignited the ascent engine, propelling them upward to dock with Gordon in the Command Module once again. For the second time, humans left the Moon behind. The LM's ascent stage dropped away, falling back onto the Moon after the docking, and the seismometers left on the lunar surface recorded vibrations from its impact that lasted for more than an hour. The trio of astronauts now spent a day taking pictures of the Moon's surface from orbit before heading for home.

On the way back, Apollo 12 marked another first when the crew witnessed an eclipse from their spacecraft windows — an eclipse of the Sun by Earth. They splashed down in the Pacific Ocean on November 24, east of American Samoa, and a camera dislodged during the final moments of the splashdown, striking Bean in the head, resulting in concussion and a brief loss of consciousness. The astronauts were recovered by the USS *Hornet*. They returned to the United States as heroes, and made a statement that the Apollo program, and repeated trips to the Moon, were here to stay.

On November 21, 1969, the Apollo 12 crew took this spectacular image of a very peculiar eclipse of the Sun, using a 16 mm motion picture camera. The Sun was disappearing behind the dark disk of our home planet as *Yankee Clipper* flew through Earth's shadow. Unfortunately the crew had run out of color film for their Hasselblad camera, forcing them to use the movie camera.

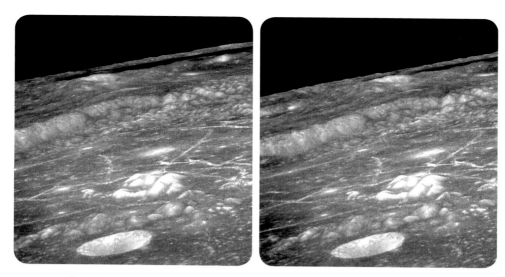

Shown here is the huge crater Humboldt, a feature spanning 189 kilometers (118 miles). The central peaks of the crater can be seen in the middle of the frame. They glow brightly because of the high angle of the Sun. Linear features and smaller craters can be seen on the floor of the crater.

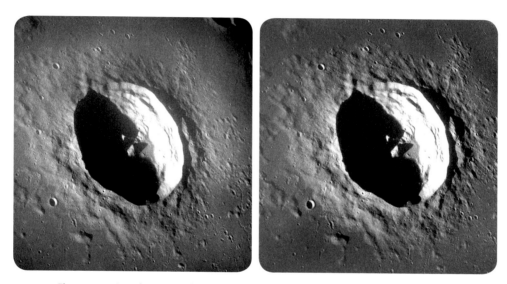

The most prominent feature near the Apollo 12 landing site was a crater called Lansberg in the Ocean of Storms some 120 kilometers (75 miles) south of the site. This lonely, circular crater spans 38 kilometers (24 miles), and features high, terraced walls rising three kilometers (1.9 miles) above the crater's floor, as well as a twin central peak. It was named for the Belgian/Dutch astronomer Philippe van Lansberge.

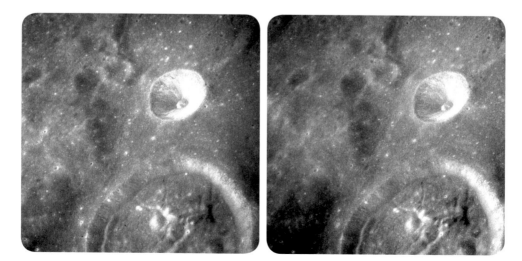

Captured from lunar orbit by the Apollo 12 crew, the pictures that make up this stereo view depict an area located east of Mare Nectaris, on the lunar near side. The prominent crater at bottom is Bohnenberger, a circular, 35-kilometer (22-mile) diameter depression with a floor featuring narrow grooves and bright central hills. The bright, bowl-shaped crater above it is Bohnenberger G. The foothills of the Montes Pyrenaeus range are visible to the left.

9 SOLVING THE "PROBLEM" (APOLLO 13)

By the spring of 1970, the Apollo program was rolling along at full speed. The first landing, and the return trip with loftier scientific goals, had been successes. Now NASA planned a third mission to the Moon that would again deepen the scientific return from the venture and increase the complexity of the experiments.

Launch for what would be termed Apollo 13 was set for April 11, 1970, using the same spacecraft configuration employed for the previous two missions. The crew for Apollo 13 consisted of Commander Jim Lovell, Command Module Pilot Jack Swigert, and Lunar Module Pilot Fred Haise. This was not the original plan, however. By tradition, the rotation would have composed a crew of Commander Gordon Cooper, Command Module Pilot Donn Eisele, and Lunar Module Pilot Edgar Mitchell. But both Cooper and Eisele were on rocky terms with NASA management at this point in their careers, for different reasons — Cooper had a relaxed attitude toward training and Eisele had failed to impress with Apollo 7 and was involved in an extramarital relationship. Because of the political issues, Deke Slayton, the Director of Flight Crew Operations, reconfigured the crew to consist of Alan Shepard, Stuart Roosa, and Edgar Mitchell. However, his superiors rejected the plan, believing that Shepard needed more time to recover from ear surgery to be flight ready.

The story becomes even more complicated. Slayton swapped the crews, moving Lovell's crew into prime position. At that time it consisted of Lovell, Command Module Pilot Ken Mattingly, and Lunar Module Pilot Haise. Mattingly, aged 34, was a Chicago-born naval aviator and aeronautical engineer who had joined the astronaut program in 1966. Three days before launch, however, NASA removed Mattingly from the mission because he had been exposed to German measles; he was replaced by Swigert. After much tumult, the final crew, hours before launch, consisted of Lovell, Swigert, and Haise.

Lovell was an experienced NASA veteran. Now aged 42, his spaceflight career included Gemini VII, Gemini XII, and Apollo 8, the first trans-lunar flight. Lovell would become the first person to make four trips into space, and the first of three people to fly to the Moon twice. Swigert, aged 38, was born in Denver and became an Air Force pilot, aeronautical engineer, and test pilot before joining the fifth group of NASA astronauts. He studied extensively and became an expert on the Apollo Command Module and requested that he become a Command Module Pilot on some mission, a request that Slayton respected. Fred Haise, aged 36, was born in Biloxi, Mississippi and became an Air Force and Marine Corps fighter pilot, a test pilot, and, like Swigert, was selected as a member of NASA's fifth group of astronauts.

With Apollo 13 set for its springtime launch, the preparations seemed normal and on schedule. The craft were named *Odyssey* (Command Module) and *Aquarius* LM. The mission's objective was to land in the highlands of the region known as

The Saturn V stack and its mobile launch tower stand atop the huge crawler-transporter that carried
the Saturn V rockets from the Vehicle Assembly Building to launch pad 39A. At the time, these tractors
were the largest self-powered land vehicles reaching speeds of only 1.6 kph (1 mph).

Fra Mauro, a region named for the 80-kilometer-diameter (50-mile) crater that lay
within the area. The region contained unusual geology and ample amounts of ejecta
that had been cast out of the impact that created nearby Mare Imbrium. This would
make for a very compelling geological area to study.

The countdown began, and the mission launched on time in the early afternoon
of April 11, lifting off high into the sky from launch pad 39A. Shortly after launch, the
second stage engine on the inboard side shut down about two minutes early, but
this was compensated for by the outboard engines burning for longer periods. The
spacecraft rose to the planned elevation of 190 kilometers (120 miles) above Earth
in a circular parking orbit. Then, some two hours later, the crew fired engines that
set the spacecraft off toward the trans-lunar injection. As this was occurring, the
crew and members of the flight crew on the ground, led by Gene Kranz, analyzed
the engine anomaly and found that it was caused by significant so-called pogo
oscillations. These vibrations caused the engine to shut down and alarmed the
ground crew; the vibrations had been observed previously during the Apollo 6 test
mission. The lessons from this flight led to modifications to avoid these oscillations

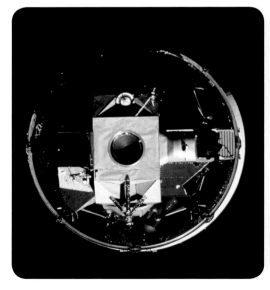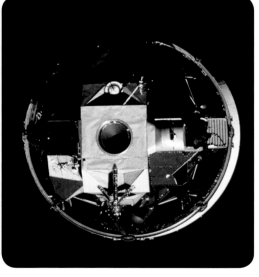

The Apollo 13 Lunar Module, *Aquarius*, is shown here still inside the Saturn third stage. The component pictures for this stereo view
were taken from the Command Module, *Odyssey*, as it approached *Aquarius* for docking and extraction from the rocket's third stage.
At the controls, Jack Swigert led the crew that was still happily on its way to the Moon, unaware of impending trouble.

There's a special poignancy to these images of Earth, at a "half-illuminated" phase, taken by the Apollo 13 crew on their way to the Moon on April 12, 1970. These were the last pictures they took of their home planet before the explosion which might have made their return to Earth impossible. Small movements in the clouds between exposures have produced some very interesting depth effects in the image.

on later flights by adding a helium gas reservoir to dampen vibrations, an automatic cutoff valve, and simplified propellant valves that the engineers believed would contain the potential problem.

After the typical separation and docking of the Command Module and the LM, the crew had the craft in the proper configuration and sent the third-stage engine on a course to impact the Moon. They cruised moonward, anticipating a smooth, three-day trip that would carry them to the interesting area of Fra Mauro.

During the cruise phase, some 56 hours after launch, the crew conducted a live TV broadcast from the capsule and then, some 6½ minutes later, began a variety of tasks. Lovell stowed away the TV camera; Haise secured the LM; Swigert was asked to perform a routine task by the ground controllers. He switched on the tank stirring fans in holding containers of hydrogen and oxygen in the Service Module. This periodic exercise mixed the gases, which were at very low temperatures, and allowed the gauges to read the tank contents more accurately. He did so, and some two minutes later the astronauts suddenly heard what they described as a "pretty loud bang." On gauges, they then observed fluctuations in the craft's electrical power and the attitude control thrusters firing by computer automation. At first, for a brief two seconds, the spacecraft lost communications and telemetry with the ground, but the computer reset the antenna to a different mode, which restored the communications.

Lovell, Swigert, and Haise were perplexed, and momentarily believed the LM had been struck by a meteoroid. Scrambling to understand the situation, Lovell uttered the immortal line, "Houston, we have a problem." Swigert confirmed this, and Lovell suggested the problem related to an electrical bus, with loss of electrical power on one of the electrical circuits. Then, the crew noticed further alarming troubles. The gauge in Oxygen Tank No. 2 read "zero." Soon thereafter, two fuel cells failed. Lovell turned around to look out of the spacecraft window and saw "some kind of a gas" leaking from the capsule out into deep space. And then troubling transformed into deeply alarming. Over the next two hours, another oxygen tank, No. 1, slowly depleted its supply of the precious gas, its gauge lagging downward to zero as well. The Service Module suddenly had no life-giving oxygen.

And if that wasn't bad enough, the situation now worsened to grave. The Command/Service Module was normally powered by creating electrical energy from the combination of hydrogen and oxygen into water; when the first oxygen tank became depleted, the only fuel cell still operating had also shut down. This now meant that the craft only had a very limited supply of reserve battery power and a small amount of water left. Scrambling, consulting with the ground controllers, the team decided to shut down and temporarily abandon the Command Module, saving it for an Earth reentry, and to crank up power inside the LM, move themselves there, and employ that portion of the craft, in their words, as "a lifeboat."

The situation was dire. The energy, electrical power, and oxygen and water were all extremely limited. The very existence of the LM and its usability as a life-support vehicle made a return trip to Earth possible; without the LM, none of the astronauts would have survived.

Once the irretrievable nature of the situation became apparent, Flight Director Gene Kranz directed personnel at Mission Control to abort the mission. This could be done in a variety of ways, as mission planners had created scenarios for anticipated in-flight problems. The most obvious alternative was the so-called Direct Abort Trajectory, which would require firing the Service Module's engines in order to spin up the spacecraft's trajectory and change its velocity by more than 6,000 feet per second. When the analysis was done, some 60 hours into the flight, it confirmed that this was a way to set the craft on a return flight toward home. However, it would only be possible at this stage if the LM were jettisoned, and this was out of the question because it now served as the only available life support for the crew. So Kranz quickly ruled out the most straightforward, anticipated emergency return option.

Kranz and the support crew on the ground also considered whether the Service Module engine's fuel could be burned off, which could be followed by jettisoning the Service Module, and then using the LM engines designed for a lunar descent to

Astronauts acting as communicators monitor the Apollo 13 emergency in Houston's Mission Control. Seated, left to right, they are: Deke Slayton, Director of Flight Crew Operations; CAPCOM Jack Lousma; and Backup Commander John Young. Standing, left to right, are Ken Mattingly and Vance Brand.

power the craft homeward. But flight controllers worried about ridding the spacecraft of the Service Module, which was needed to supply valuable heat shield protection during Earth reentry. Kranz and his colleagues were also worried about the structural integrity of the Service Module, and whether its engines would be safe to use in a rescue scenario, so they wished to avoid employing them if at all possible.

Moreover, the spacecraft was approaching the point where a "breakeven" in terms of energy would exist between a direct return to Earth and a circumlunar return — flying around the Moon and slingshotting back toward Earth. Ground controllers decided to abandon the idea of the direct return scenario in favor of a circumlunar approach. This would also give them more time to plan how the return would safely get the astronauts back to Earth.

The circumlunar option would use the Moon's gravity to propel the spacecraft home. Because the crew had been instructed to chart a course that would get them to the region of Fra Mauro, the craft had departed from its initial trajectory prior to the accident. So in order to initiate a free return trajectory that would carry them on the proper orbit, they needed to conduct a short burn with the LM descent engine. This burn lasted about 30 seconds.

Inside *Aquarius*, conditions worsened for the crew as the minutes and hours ticked by. The supply of oxygen was critically low, and would be marginally enough to see the crew back to Earth. On the ground, Kranz plodded along with his team, enlisting Ken Mattingly to help controller John Aaron to devise a plan so that *Odyssey* could be restarted, facilitating the landing on Earth. In Houston, Kranz simply declared, "Failure is not an option."

As *Aquarius* approached the Moon, the astronauts were increasingly uncomfortable. They looked down upon the lunar surface, coming as close as 254 kilometers (137 miles) to the Moon at closest approach on April 15. Lovell gazed onto the surface and lamented the fact that he and Haise would not walk on the Moon. The astronauts then focused on the business of returning to Earth, their close approach to the Moon now history. The temperature inside *Aquarius* was near freezing, electrical power was dangerously low, and Fred Haise began to feel increasingly sick. He was suffering from a urinary infection and running a fever, very low in energy and feeling terrible. It was a risky and potentially calamitous situation. The lifeboat was becoming a very difficult place in which to be alive.

Apollo 13's Lunar Module, *Aquarius*, imaged as it was jettisoned a few minutes before 11 a.m. CST on April 17, 1970. This was just over an hour before the Command Module's splashdown in the South Pacific Ocean. *Aquarius* had acted as a lifeboat for the imperiled astronauts.

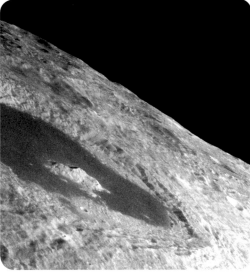

The Apollo 13 crew captured this breathtaking image of the crater Tsiolkovsky on April 14, 1970, as they swung around the lunar far side. This prominent feature, never visible from Earth, was named after the famous Soviet engineer Konstantin Tsiolkovsky, following its identification in *Luna* 3 images. It is 180 kilometers (113 miles) across, with a central peak rising 3,400 meters (11,050 feet) above the floor.

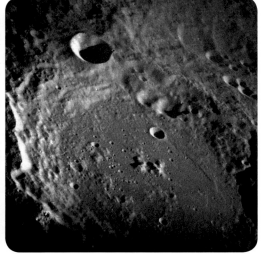

The far side crater Chaplygin was photographed by the Apollo 13 crew, as their craft came around the Moon, heading straight back toward Earth. Chaplygin is a 123-km (787-mile) wide crater with terraced walls, a smooth floor, and a central peak.

The New York Times of April 15, 1970, reflecting the massive worldwide public concern about the dangers the Apollo 13 crewmembers faced.

The psychological state of the crewmembers also began to crumble slightly. Swigert suspected that the craft would not be able to return to Earth and that the flight controllers were withholding that information from the crew. Haise, feverish and very ill, felt that Swigert's limited experience may have played a role in causing the accident. Before a full-out argument could get rolling, however, Lovell refocused his colleagues on the mission of getting home. (When I interviewed Jim Lovell many years later, he did tell me that the notion that the three of them would become the first "human popsicles in permanent orbit" was weighing very heavily on all of their minds.)

Following their closest approach to the Moon's surface, the crew conducted another burn of the LM descent engine. This had the critical effect of speeding up the return velocity to Earth by about ten hours, and also shifting the landing spot from the Indian Ocean to the Pacific Ocean. Kranz and the other controllers felt that the crew, despite the deteriorating conditions, would have sufficient oxygen, electricity, and water to return to Earth on this timetable, avoiding a riskier maneuver that would have involved jettisoning the Service Module. After this key burn of the LM descent engine, which lasted 4½ minutes, only two small course corrections were made before approaching the home planet.

The grave circumstances of Apollo 13 became a worldwide sensation. Television coverage broke in and reported routinely on the mission in a special report scenario that harkened back to the assassination of John Kennedy and other tragedies. Updates did not look particularly encouraging for some time, leading to a sense of panic and a feeling that a terrible tragedy was imminent. The small supply of electrical power in the craft did not allow live television broadcasts, so news desks passed on the latest statements and updates as broadcasters employed models of the spacecraft to try to explain to viewers what was happening and how the mission might end.

As Lovell later recounted, the LM, used as a lifeboat for the three astronauts, was really stretched to the limits of what it was designed to do. It was supposed to sustain two astronauts for about a day and a half, rather than three astronauts for the four-day return to Earth. The Command/Service Module produced water as a byproduct from its fuel cells. But such was not the case with the LM, which employed silver-zinc batteries, and so within the LM, electrical power and water were vitally important and in short supply. Oxygen was less of a concern because the LM had enough in order to repressurize the spacecraft's atmosphere after each lunar surface activity. But Kranz and the astronauts all wanted to minimize the inherent risks of course, so they powered down the LM as much as possible to preserve supplies and to keep the communications capability and life-support functions as stable as possible until nearing reentry in Earth's atmosphere.

Another looming problem that worsened as the cruise toward Earth continued was carbon dioxide. This, in fact, proved to be one of the most difficult challenges for the controllers and the crew. Removing carbon dioxide from the LM's atmosphere would be critical — too much buildup of this gas would prove fatal. Carbon dioxide removal normally took place via lithium hydroxide canisters and the LM now lacked the number required to remove sufficient carbon dioxide for the return cruise home. Further, the backups were in storage, out of reach, and the Command/Service Module lithium hydroxide canisters were incompatible with the LM. So ground controllers worked out a plan, communicated it to the crew, and the astronauts improvised a system to accomplish the task, using hoses to connect the square section Command Module canisters to the LM's cylindrical sockets. They later referred to this arrangement as "the mailbox."

Additionally, as the craft approached Earth, the crew would have to power up the Command Module from scratch. This had never before been tried during a flight. Given the craft's weak power availability, controllers Aaron and Mattingly proposed a procedure, but they were uncertain. The extremely cold temperatures within the spacecraft, which dropped as low as 4° C (39° F), meant that water condensed on surfaces and alarmed the crew and controllers, who anticipated potential electrical shorts. But this did not, in the end, turn out to be a problem.

Another risky maneuver now had to be negotiated. Before approaching Earth's atmosphere, the crew would have to separate the LM from the Command Module. Ordinarily, the astronauts would employ the Service Module's so-called reaction control system, with its thrusters, to accomplish this. But the craft's power failure meant this was not operable, and the Service Module would be gone before the LM anyway. University of Toronto engineers, led by Bernard Etkin and called on by Grumman, worked on the problem for a day and proposed pressurizing the tunnel connecting the LM and the Command Module just prior to reentry. This, they proposed, would push the two craft away from each other. They communicated their slide-rule calculations to Mission Control, who sent them on to the astronauts. The procedure worked.

With Apollo 13 approaching home, the world watched with bated breath. The sustained crisis brought people of all nations together and the world prayed for a safe return. News reports became ever more frantic, and despite the constant analysis, all that could now be done was to wait and hope.

During the Apollo 13 emergency, Deke Slayton (standing at left) explains a proposed procedure for constructing lithium hydroxide canisters to remove excess carbon dioxide from the Lunar Module cabin. This was the biggest threat to the astronauts' well-being. Members of the operations team listened, from left to right: Howard Tindall, Sigurd Sjoberg, Chris Kraft, and Robert Gilruth.

As the crew moved toward Earth, they first pulled away from the Service Module, using the LM's thrusters, and photographed the damage on the Service Module as they inched away from it. They were stunned to discover that an entire panel on the side of the Service Module was missing. Safely back within the Command Module, the crew then jettisoned the LM, and, now trusting their fate to *Odyssey*, began a reentry into Earth's atmosphere. The normal communications blackout period for a returning Apollo spacecraft was about four minutes. The blackout for Apollo 13 lasted six minutes, longer than expected, and greatly heightened the drama on the ground.

But then, in a flash, communications returned, the craft was spotted, and *Odyssey* made a splashdown, its chutes successfully deployed, in the Pacific Ocean, southeast of American Samoa. The amphibious assault ship USS *Iwo Jima* picked up the crew, and the world celebrated. Lovell and Swigert were in reasonably good shape, and Haise would recover from his infection and fever.

In the end, the most dangerous circumstance in space exploration ended with a triumph. The astronauts missed walking on the Moon, but they lived to tell their tale and walked again on planet Earth.

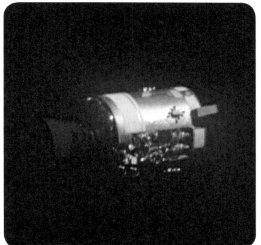 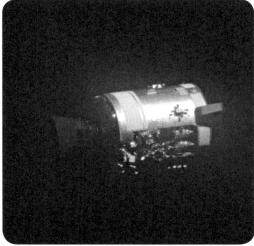

On April 17, 1970, a few hours prior to the splashdown in the South Pacific, Apollo 13 astronauts shot a short sequence of images of the jettisoned Service Module from the Command Module, *Odyssey*. We used two consecutive frames from this series to construct a stereo view, showing the damage from the oxygen tank's explosion.

Another view of the damaged Service Module taken prior to splashdown, clearly showing what had happened: an entire panel (top) had been blown off by the oxygen tank explosion, and the interior damage was substantial. In this view the S-band antenna is visible above the damaged area, and on the right side are the service propulsion system engine and nozzle.

Following the successful splashdown of Apollo 13, NASA flight directors celebrate in Mission Control. They are, left to right: Gerry Griffin, Gene Kranz, and Glynn Lunney.

Rear Admiral Donald Davis (at right) welcomes the Apollo 13 astronauts aboard the recovery ship USS *Iwo Jima*. From the left: Fred Haise, Jack Swigert, and Jim Lovell.

This image provides a sea-level view of the recovery of Apollo 13 in the South Pacific. Fred Haise steps onto the life raft as Jim Lovell leaves the spacecraft in the background. Jack Swigert is already in the raft. The crew were picked up by helicopter and carried to the waiting USS *Iwo Jima*.

Jim Lovell reading a newspaper's account of his safe return to Earth.

The crew of Apollo 14 pose in front of their mission badge at the Kennedy Space Center during training. They are Command Module Pilot Stuart Roosa, Commander Alan Shepard, and Lunar Module Pilot Ed Mitchell.

10 EXTENDED EXPLORATION
(APOLLO 14, SOYUZ)

In the wake of the near disaster with Apollo 13, NASA made precautionary plans with the following mission to tighten safety regulations and procedures. The eighth manned mission in the Apollo program, Apollo 14, was scheduled for a liftoff in late January 1971. In the United States, President Richard Nixon was still grappling with the Vietnam War, declaring that the "Vietnamization" of the war was in full swing and that the combat mission of American troops would end by the coming summer. The administration's criminal activities that would result in the Watergate scandal were underway but as yet undetected. The cultural leaders of the rock 'n' roll movement, the Beatles, had now broken up, leaving a wide open and uncertain future for the leading edge of what was then called "pop" music, but would now be classified as "rock." The pure idealism of the sixties now seemed faded; the hippie culture had subsided, and although no one quite knew it yet, the "me decade" of self-interest was already rolling forward. In the Soviet Union, the space program kept moving, but the momentum for a manned lunar program was now completely gone.

For the first Apollo mission of 1971, NASA turned to a wily veteran. Alan Shepard had been the first American in space, making his suborbital flight May 5, 1961, just 20 days before John Kennedy's speech calling for a mission to land on the Moon. Shepard, now 47, was born in New Hampshire and had a distinguished career as a naval aviator and test pilot before his Mercury flight in *Freedom 7*. Now he was slated to be the Commander of Apollo 14, and this would make him the oldest person ever to walk on the Moon, as well as the only Mercury astronaut to accomplish this feat.

Joining Shepard in this crew were Command Module Pilot Stuart Roosa, and Lunar Module Pilot Edgar Mitchell. Shepard's spaceflight experience would be especially valuable: both Roosa and Mitchell were rookies, having never yet flown in space. Roosa was an aeronautical engineer, Air Force pilot, and test pilot. Age 37, he was born in Colorado and achieved an impressive military record before being chosen as one of NASA's astronaut class of 1966. Mitchell, aged 40, was a Navy officer and aviator, test pilot, and aeronautical engineer. Born in Texas, he was also selected in the 1966 astronaut group and had served in support teams on previous Apollo missions before his assignment to Apollo 14. The mission's backup crew consisted of Gene Cernan, Ronald Evans, and Joe Engle. With Harrison Schmitt substituting for Engle, this crew would become the primary one for the last Apollo mission. This time, the Command/Service Module was nicknamed *Kitty Hawk*, and the Lunar Module *Antares*.

Following Apollo 13, NASA engineers modified the electrical power system of the Service Module, attempting to minimize the risk of any further malfunctions. The team redesigned the oxygen tank in which a spark had occurred and caused the explosion when Swigert switched on the stirring fans. They added a third tank too. Confidence in the new design was high.

In lunar orbit, the Apollo 14 Command Module, *Kitty Hawk*, floats in this image taken from the Lunar Module, *Antares*. Alan Shepard and Edgar Mitchell commenced their descent to the lunar surface, while Stuart Roosa stayed within the orbiting Command Module.

The launch date for Apollo 14 was originally set for October 1, 1970. But the Apollo 13 experience changed the timetable, pushing it back. The new launch date for Apollo 14 was scheduled for January 31, 1971, and the mission was to aim again at the region of Fra Mauro, the area targeted by the aborted Apollo 13. This region of highlands, named after the crater lying within it, consists largely of ejecta from the immense impact that created the nearby Mare Imbrium. Studying this hilly geological area would provide insights into the formation of Mare Imbrium and also allow the study of large amounts of debris covering the ejecta thought to be exposed older rocks from deep below. This might allow the explorers to uncover some secrets about the Moon's geological history. Now the Apollo missions were evolving from simple exploration and wonder at just being on the lunar surface to a deeper and more organized program of scientific studies.

The January launch took place right on schedule despite heavy cloud cover that hung over Kennedy Space Center. With US Vice President Spiro Agnew on hand, along with Spanish Prince Juan Carlos and his wife, Princess Sophia, the Saturn V jumped skyward and quickly out of sight due to the cloud cover. The spacecraft achieved Earth orbit and then Shepard separated the Command/Service Module from the Lunar Module and turned it around for docking.

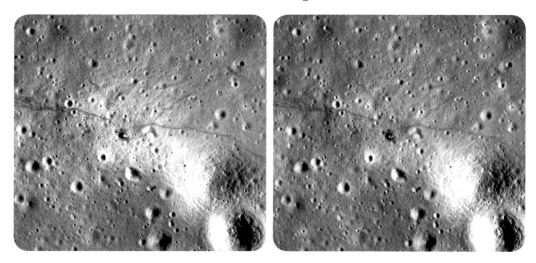

The Apollo 14 landing site in the highlands of Fra Mauro appears in this image made with the Lunar Reconnaissance Orbiter on November 28, 2009. The descent stage of the LM *Antares* is clearly visible, as are tracks made by Alan Shepard and Edgar Mitchell with the small wheeled cart they used to transport equipment and specimens. The width of the image is about 350 meters (1,138 feet).

And then the mission had its first hint of trouble. The astronauts worked the docking procedure multiple times, having trouble each time completing the maneuver. Finally, after more than 1½ hours of trying, Roosa tried pushing *Kitty Hawk* with its thrusters against *Antares* and the docking latches took hold, accomplishing the procedure. At the time, it seemed like a close call and potentially a major problem.

On February 4, Apollo 14 concluded its glide phase over the 386,000-kilometer (240,000-mile) trip to the Moon. Entering lunar orbit, the spacecraft again seemed fine. The following day, Shepard and Mitchell climbed into the LM and prepared for their descent to Fra Mauro. Roosa would stay within *Kitty Hawk*, piloting it as it circled the Moon. Soon after the beginning of the descent, within *Antares* the astronauts encountered a problem. The lander's computer signaled an "abort" alert, and the astronauts determined that it was a false alarm due to a faulty switch. But if the alarm were to recur after the descent engine began firing, the computer would treat the false alarm as if it were real and abort the descent. This would cause the craft's ascent stage to reignite and separate from the descent stage, and the LM would return to a lunar orbit.

Back at Mission Control, the flight team again enlisted engineers, at NASA and at the Massachusetts Institute of Technology, to work on the problem. After a short time, engineers suggested reprogramming the computer onboard *Antares* to ignore the abort signal. Mitchell frantically entered the changes into the computer, and it worked, allowing the descent to begin. "It's a beautiful day to land at Fra Mauro," said Shepard in response to the fix.

But another problem cropped up. The landing radar employed by *Antares* failed to recognize the lunar surface, so that altitude and vertical speed data would not show in the LM. The fix this time seemed to be in cycling through the craft's radar breakers. At an altitude of about 5,500 meters (18,000 feet), the data readouts came back on, allowing the astronauts to safely pursue the landing. The spacecraft pitched over and Shepard and Mitchell began to see landmarks on the Moon. "There it is," said Shepard of Fra Mauro, and he manually landed the LM. "It's really a wild looking place here," said Mitchell. And the craft came to a halt, just where they had planned. In fact, Shepard's landing came closer to the chosen point than any of the other five lunar landings.

On February 5, Shepard and Mitchell planned their first of two moonwalks, which would last between 4½ and 5 hours each. They named the lunar base at their

Edgar Mitchell operates a TV camera on the cratered soil of Fra Mauro, casting a very long shadow. Astronauts have reported that shadows on the Moon can look quite alien; they are extremely dark due to the absence of an atmosphere to scatter light. This image was shot by Alan Shepard.

One of the first tasks for Alan Shepard and Edgar Mitchell at their Fra Mauro Base was to deploy the S-band antenna. This folding dish, spanning three meters (9.75 feet) and with a tripod standing 1.5 meters (4.9 feet) tall, provided an improved television signal for transmission to Earth. The shadow of the LM, *Antares*, is also visible on the left side of this image, as well as part of the flag.

In this view, the Apollo 14 LM, *Antares*, reflects a circular flare caused by brilliant sunlight over the Fra Mauro highlands. Shepard and Mitchell said the unusual ball of light had a jewel-like appearance. At the extreme left of the image is the lower slope of Cone Crater.

landing position Fra Mauro Base, which was subsequently added to lunar maps showing the Fra Mauro region. As he descended the LM ladder and finally touched the lunar surface with his heavy boot, Shepard said, "And it's been a long way, but we're here." This was the third "first step" of a lunar explorer on a new mission, the first two steps coming from Neil Armstrong and Pete Conrad, and Shepard waited a short time and stepped away from the ladder before uttering his historic line.

This time, the astronauts had succeeded in bringing along and employing a color television camera, which they planted on the surface at Fra Mauro Base, along with the customary US flag. So there would now be broadcasts in natural color showing the astronauts in their Moon walks and activities. Shepard wore an Apollo suit that had red stripes on the arms and legs, enabling easy identification of the Commander, as opposed to Mitchell. NASA had not been satisfied with the appearance of the Apollo 12 pictures, in which it is very difficult to differentiate between Conrad and Bean. Enjoying this easy identification, NASA continued the practice with the remaining Apollo flights and on into the era of the Space Shuttles. As with the previous missions on the lunar surface, the astronauts deployed the ALSEP, the Apollo Lunar Surface Experiments Package, which contained experiments that would record data on seismology, magnetism, the solar wind, heat flow, and detection of ions. They also deployed a pull-cart that would be used for transporting equipment and Moon rocks, the Modular Equipment Transporter (MET). The astronauts nicknamed the cart the "lunar rickshaw."

During the Apollo 14 mission, Alan Shepard stands by the Modular Equipment Transporter, a wheeled cart made for carrying tools, cameras, and sample cases over the lunar surface. Shepard has a vertical stripe on his helmet to identify him as the Commander.

Here, Edgar Mitchell sets up a suite of scientific experiments, called the ALSEP package, near the Apollo 14 LM landing site. The photo was taken by Alan Shepard and features his broad shadow at bottom. The instruments were designed for long-term study of various lunar phenomena.

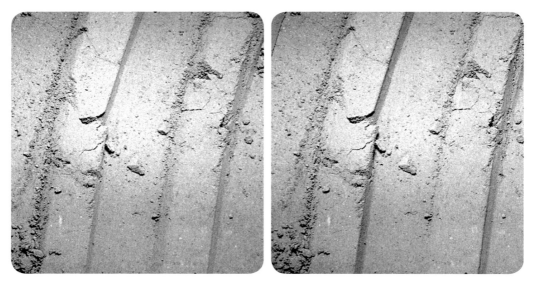

On the second Apollo 14 moonwalk, Alan Shepard made this image of an astronaut boot print using the ALSCC. The lunar soil retains the shape of the boot print, including the nearly vertical sides of the treads, because it is slightly cohesive. The lunar soil's "stickiness" results from many factors, including particle size and electrostatic charge.

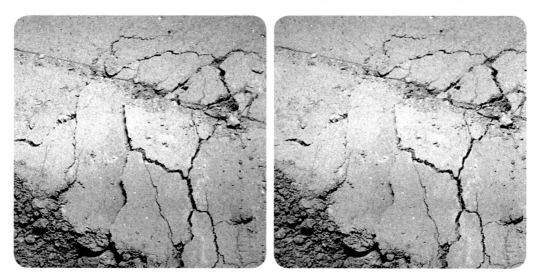

Tire tracks left in the soil by the Modular Equipment Transporter (MET), nicknamed the "rickshaw". Cracked and uplifted blocks again indicate adhesion between soil particles.

Setting off with the lunar rickshaw, Shepard and Mitchell began the traverse for their first moonwalk. They established the TV camera in a way such that viewers at home could watch a good portion of their journey, although the two ducked out of view on occasion as they walked into areas of low terrain. Their journey took longer than expected, with Mitchell carrying the ALSEP package, but he remarked that the trip was "unexpected" more so than "difficult." The rolling terrain seemed very different than the flat surface they had trained on. Once they arrived at the ALSEP site, the stiff gloves and suits made deployment clunky, which caused Mitchell to say, "It takes two of us to do what half of us can [normally] do!" They ran about half an hour behind schedule on the first walk, but encountered no substantial problems. Anxious to start a second walk and to make the most of it, the two astronauts talked ground controllers in Houston into beginning the second walk early. Sleeplessness was a factor: once stretched out in their hammocks inside the LM, the tilt of the craft seemed exaggerated, and they awoke several times with the strange feeling that the LM was tipping over. In the end, they got about four hours of sleep each and were eager to set off in exploration mode. The first moonwalk lasted nearly four hours 48 minutes and succeeded in all the astronauts hoped to accomplish.

So some 13 hours after the first walk ended, the astronauts commenced their second extravehicular activity. Shepard again set foot onto the lunar surface first, followed by Mitchell some seven minutes later. During the second walk, the astronauts planned to walk to Cone Crater, a 305-meter (1,000-foot) wide depression in Fra Mauro.

The rolling terrain surrounding the LM continued for a substantial distance to the east. They commenced by moving some 200 meters (650 feet), and stopped to take pictures, measure the Moon's magnetic field, and collect samples. After finishing at what was dubbed Station A, they continued onward and after an hour had elapsed, got to a point 650 meters (2,130 feet) away from the LM. There they noticed they had crossed onto the ejecta blanket from Cone Crater.

With the grade angling slightly uphill, Mitchell noted that the walk was a little more exerting. "We're starting uphill now," he said. "It's definitely uphill." Shepard and Mitchell stopped short of the crater by about 30 meters (100 feet) and collected a substantial amount of lunar rock and soil samples. These would be of particular interest geologically, it was believed, because they had been blasted up from deep below the surface.

The pair continued walking along and Shepard noticed how dusty the lunar surface was – how dust was getting kicked up and adhered to the spacesuits. "Nothing like being up to your arms in lunar dust," said Shepard.

The two were determined to make it to the rim of Cone Crater, across this uphill grade. Shepard moved ahead, pulling the cart, as Mitchell followed and narrated a monologue of what they were seeing for the colleagues in Houston. The steep gradient made the narrative hard to decipher back at Mission Control. On the ground, as breathing intensified and the heart rates of the astronauts rose, Fred Haise called for the two to stop and rest. Houston suggested that they turn back. Mitchell argued they should continue. Shepard intervened and the two kept going, but they ultimately shot past the rim of the crater, moving 30 meters (100 feet) beyond it, before heading back.

The second moonwalk lasted nearly four hours 35 minutes. Near its end, Shepard made a surprise announcement toward the color TV camera that was running. He stepped toward it and pulled out a makeshift golf club. "Houston, you might recognize what I have in my hand as the handle for the contingency sample return," he blurted out. "It just so happens to have a genuine six iron on the bottom of it. In my left hand, I have a little white pellet that's familiar to millions of Americans."

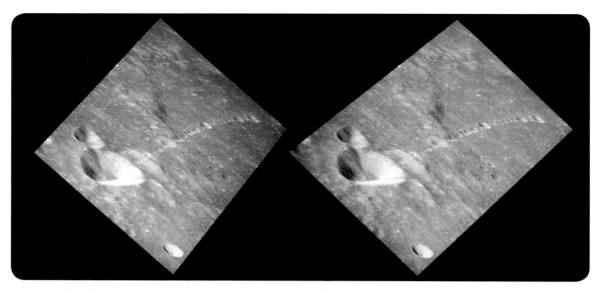

The Davy crater chain is an unusual lunar feature that was originally a candidate for the Apollo 14 landing site. Consisting of 23 small craters, it stretches for 47 kilometers (29 miles). Long ago thought to be volcanic, it's now known to be an impact feature, created by an object – a meteoroid or cometary nucleus – that fragmented before impact.

Shepard plopped a golf ball onto the lunar surface. He was an avid golfer and had planned this surprise as a test of his own abilities on the Moon's surface and also as a test of the weaker lunar gravity, relative to Earth. "Unfortunately, the suit is so stiff, I can't do this with two hands," he said. "But I'm going to try a little sand trap shot here."

The Apollo 14 Commander took a swing, knocking some lunar dust skyward. "Hey, you got more dirt than ball," said Mitchell. Watching from back on Earth, Fred Haise, recovered and acting as a CAPCOM on the ground, said, "That looked like a slice to me, Al." "Here we go again," said Shepard as he swung again. "Straight as a dime. Miles and miles and miles." Shepard later concluded that the golf ball traveled between 180 and 370 meters (200 to 400 yards). Not to be outdone, Mitchell thrust a lunar scoop handle in the air as if it were a javel n, in what could perhaps be called a Micro Lunar Olympics.

They concluded the second moonwalk after collecting some 43 kilograms (94 pounds) of Moon rocks. After 33½ hours on the Moon and 9½ hours walking around, the pair prepared to blast off and return to *Kitty Hawk*, with Roosa still orbiting overhead.

On February 9, 1971, the Apollo 14 crew splashed down in the South Pacific and were recovered by US Navy personnel dispatched from the USS *New Orleans*. In this image Edgar Mitchell steps out of the spacecraft and Shepard and Roosa follow.

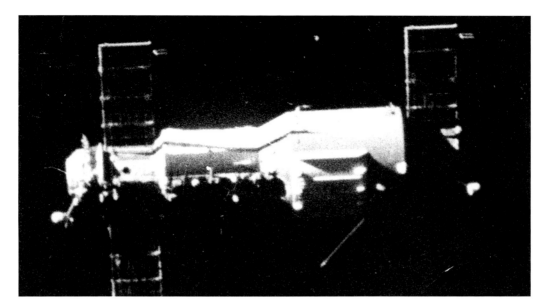

The Soviet Salyut 1 space station floats in low-Earth orbit as it is photographed from the departing Soyuz 11, in 1971.

On their return cruise back toward Earth, the astronauts conducted the first materials processing experiments ever done in space. *Kitty Hawk* splashed down in the Pacific Ocean, south of American Samoa, and the astronauts recovered well from the ordeal. In the wake of the scare over Apollo 13, the Apollo program was now solidly back on the right track, and the samples collected and science that had been conducted would help the Moon mission enormously.

In the Soviet Union, their ongoing space program now essentially consisted of watching the Americans run away with the lunar prize. The Soyuz program continued, but these flights were Earth orbital missions. Through 1970, eight Soyuz missions had taken place, all orbiting our planet, some with rendezvous objectives and others paving the way for coordination with the Salyut Space Station. This was the world's first space station, named Salyut 1, to be deployed by the spring of 1971. The launch of this very significant "first" was planned to mark the 10th anniversary of Yuri Gagarin's flight, but in the end had to be delayed by several days.

The crew of Soyuz 11 pose before their 1971 flight. From left to right: Georgi Dobrovolski, Vladislav Volkov, and Viktor Patsayev, which constituted the first group of men to occupy a true space station. Tragically, the crew died in space from asphyxiation.

The first mission to Salyut 1 was Soyuz 10, launched on April 22, 1971. The crew, Vladimir Shatalov, Aleksei Yeliseyev, and Nikolai Rukavishnikov, hoped to rendezvous with Salyut 1 and board the space station, becoming the first station crew in space exploration history. But the docking with Salyut 1 was unsuccessful and the crew had to return to Earth.

Several months later, another crew set off for Salyut 1. Soyuz 11, with cosmonauts Georgy Dobrovolsky, Vladislav Volkov, and Viktor Patsayev, blasted off from Baikonur on June 6 and docked with Salyut 1 a day later. The cosmonauts remained on board the space station for 22 days, setting an endurance record in space that would stand for another two years. When they entered the station, the crew discovered a smoky smelling atmosphere. On the 11ᵗʰ day in the station, a small fire broke out. The cosmonauts hoped to observe a rocket launch from the station, but the launch was delayed. They broadcast TV back to Earth and generally enjoyed the stay.

Tragically, on June 30, when the capsule was recovered on Earth after landing, the mission recovery team opened it to find all three crewmembers dead. The cosmonauts had bluish patches on their faces and had hemorrhaged from their mouths and noses. The official cause of death was given as asphyxiation. The original Commander of the mission, Alexei Leonov, had advised the crew to close valves between the orbital and descent modules manually, because he did not trust the automatic mechanism. The cosmonauts did not follow his suggestion, however, and this may have been the cause of the fatal event. Once again, a Soviet mission had ended in a shocking disaster.

With respect to the Moon, the Soviets did carry on with the Lunokhod program, a meticulously planned series of missions to place *unmanned* rovers on the Moon. Russian for "moonwalker," the Lunokhod was intended to provide a backup and support capability for the later Soviet manned missions to the Moon. After years of testing in secret, the first Lunokhod launched in 1969, but shortly after launch the rocket disintegrated. As a follow-up, Soviet engineers designed Lunokhod 1,

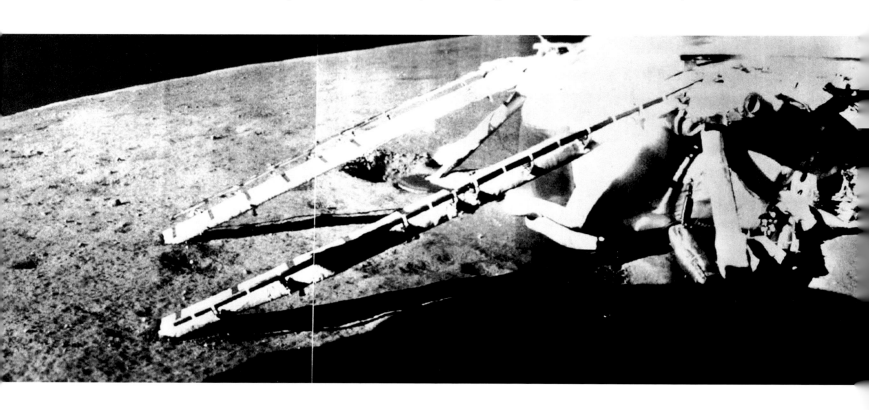

A replica of Lunokhod 2 photographed in stereo at the NASA exhibition *A Human Adventure*. Lunokhod 2 weighed 840 kilograms (1,852 pounds) on Earth, and was equipped with two vidicon television cameras for navigational control. The images were used by a five-man team of controllers on Earth to drive the rover in real time.

which launched on November 10, 1970. About 2.3 meters (7.7 feet) long, the rover was similar in appearance to later Mars rovers that have explored the Red Planet. Its eight 'skeleton' design wheels ferried it around, and it carried a suite of instruments, including antennas, television cameras, soil testing devices, spectrometers, an X-ray telescope, and a cosmic ray detector. Lunokhod 1 successfully landed on the Moon in Mare Imbrium on November 17 and operated its experiments until September 1971. A second, more complex Lunokhod was developed and launched to the Moon in 1973.

The Soviet program, although it would not land men on the Moon, made important contributions to understanding lunar science. And underneath the veil of the space race, something else was happening. Alongside the direct competition, Soviet and American space explorers were forming a brotherhood that would become stronger and stronger as the months and years passed.

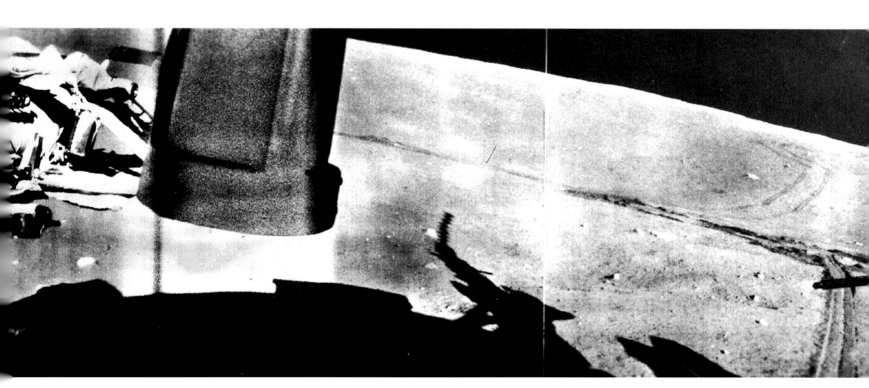

The Soviet space program reached the Moon in 1970 with the first of two Lunokhod rovers, robotic explorers that pioneered later rover technology used for Mars missions. This image shows the ramp used to roll the rover off the lander onto the lunar surface and a panorama of the barren lunar landscape.

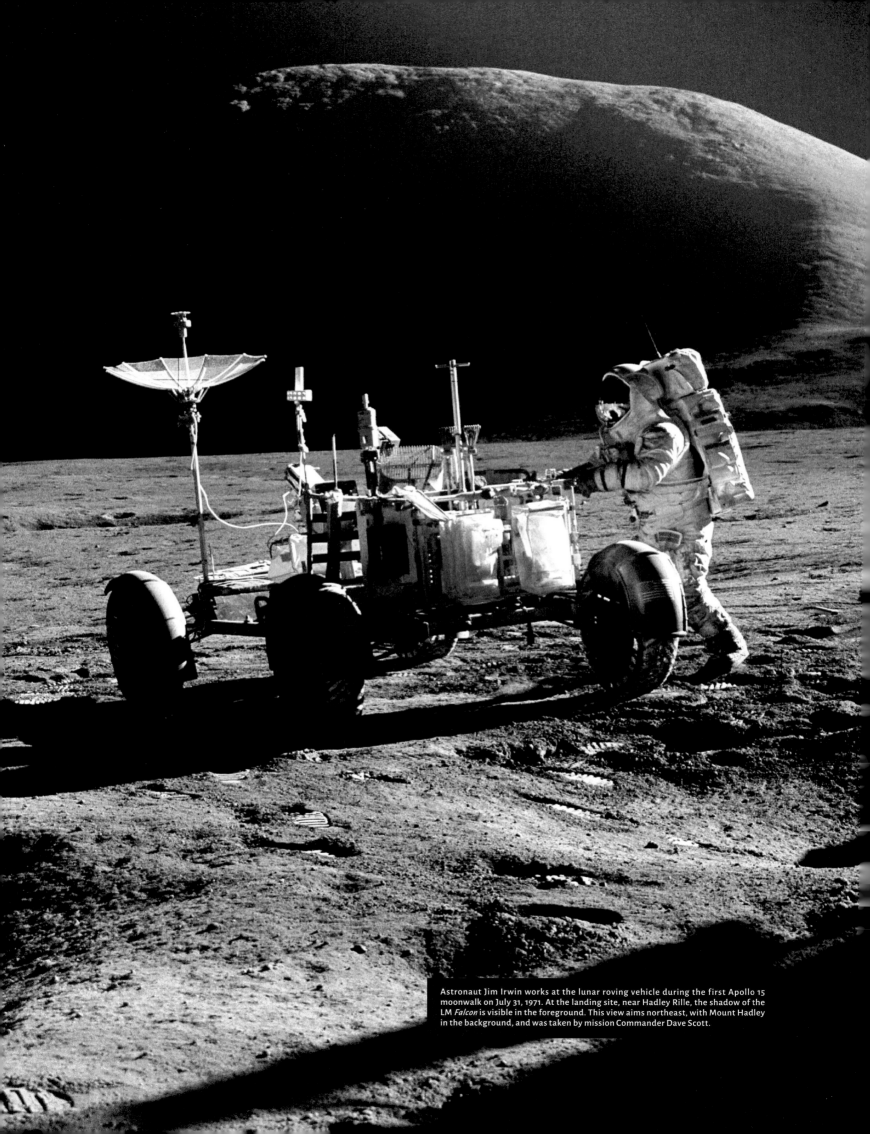

Astronaut Jim Irwin works at the lunar roving vehicle during the first Apollo 15 moonwalk on July 31, 1971. At the landing site, near Hadley Rille, the shadow of the LM *Falcon* is visible in the foreground. This view aims northeast, with Mount Hadley in the background, and was taken by mission Commander Dave Scott.

11 ENTER LUNAR ROVERS
(APOLLO 15)

Flushed with the success of Apollo 14 and with the near-tragedy of Apollo 13 now firmly in the past, Apollo's mission planners expanded the length and sophistication of the next mission. Apollo 15 was designed to be the first of the so-called J missions, a chapter in NASA's playbook that designated longer stays on the lunar surface with an expanded set of scientific goals.

To everyone's excitement, the mission was also the first to carry on board a *lunar roving vehicle*, the world's most expensive car! With this rover, the amount of lunar territory that could be explored was greatly expanded since the astronauts were no longer limited to walking in the bulky spacesuits to various destinations. The lunar rover made driving to areas on the Moon classy and comfortable; it also enabled the expansion of scientific studies and the collecting of Moon rock samples from a larger area than was previously possible. The Apollo 14 mission achieved its aims with no significant problems. NASA was almost making space travel look easy.

Following Apollo 14 after a six-month period of preparation, Apollo 15 was scheduled for launch on July 26, 1971. The sweeping social and psychological change that marked the end of the sixties continued on apace. In Los Angeles, Charles Manson and his followers were found guilty of the Tate–LaBianca murders, and just a few days later a major earthquake rocked the area. In the spring, half a million people marched in Washington to protest the Vietnam War, and the *New York Times* began to publish the Pentagon Papers, a classified study of the situation in Vietnam. In New York, workers completed the top (110th) floor of the South Tower of the World Trade Center: this edifice and its sister the North Tower – together named the World Trade Center, but popularly known as the *Twin Towers*, became for a few months the tallest buildings in the world, and a symbol of America's prosperity.

The world of music and pop culture also raced forward in rapid-fire change. With Jimi Hendrix and Janis Joplin gone, victims of excess, Jim Morrison of The Doors would follow shortly during the summer of 1971. The Beatles no longer toured and were now experimenting with solo careers. A new heavy rock wave of British music invaded the United States, spearheaded by the New Yardbirds, who swiftly transformed into the phenomenon of Led Zeppelin. In London, members of a new rock band, formed in 1970, were writing and recording their first album. The band was called Queen and consisted of Freddie Mercury, Brian May, John Deacon, and Roger Taylor. With that line-up, the band played its first show outside London less than a month before the launch of Apollo 15.

The Apollo 15 crew consisted of one experienced spaceflight veteran and two rookies. Dave Scott, the Commander, was a 39-year-old Air Force officer and test pilot. Texas born, Scott had been chosen as one of NASA's third astronaut group and subsequently teamed up with Neil Armstrong as part of the Gemini VIII mission.

He then served as Command Module Pilot on Apollo 9, alongside James McDivitt and Rusty Schweickart. Scott's experience would be key for the ambitious agenda prepared for Apollo 15. He was joined by Command Module Pilot Al Worden, a 39-year-old Michigan-born man who had been an Air Force pilot and was selected as an astronaut candidate in 1966. The mission's Lunar Module Pilot was Jim Irwin, aged 41, who had a background as an aeronautical engineer and test pilot as well as being an Air Force pilot. Irwin was Pittsburgh born, selected in the 1966 group of astronauts along with Worden.

The crew's backups were Commander Dick Gordon, Command Module Pilot Vance Brand, and Lunar Module Pilot Jack Schmitt.

The design for Apollo 15 was ambitious and exciting. The target area near Hadley Rille is a section of the lunar surface marked by flexure, appearing as a groove cut into the lunar floor. Nearby stands Mons Hadley, a mountainous rise that reaches more than 1,360 meters (4,465 feet) above the floor below and has a diameter of about 25 kilometers (15 miles). Hadley Rille lies within an area of the Moon known as Mare Imbrium, the Sea of Rains. On the other side of the landing site from Hadley Rille are the lunar Apennine Mountains, a rugged range forming the southeastern rim of Mare Imbrium. The specific area where the spacecraft was to land was dubbed Palus Putredinus, the Marsh of Decay.

Rather than spending roughly a day on the lunar surface, as had Apollos 11, 12, and 14, this fourth mission to explore the Moon with humans would allow three days of investigation. Worden would stay within the Command Module, nicknamed *Endeavour*, in lunar orbit. Scott and Irwin would descend to the surface in the Lunar Module, with its callsign *Falcon*, to explore the Moon, conduct scientific experiments, and collect Moon rocks, all achieved by driving around in a Moon rover.

The lunar rover was the result of a complex engineering project that had been in development for years. Manufactured by Boeing and General Motors, the rover quickly acquired the name "Moon buggy" in a play on "dune buggy." It was battery

The Apollo 15 crew trained extensively prior to the mission and are seen here aboard the NASA Motor Vessel *Retriever*, speaking with the lead Underwater Demolition Team swimmer as they practiced in the Gulf of Mexico. Posing here, left to right, are Fred Schmidt along with astronauts Al Worden, Jim Irwin, and Dave Scott.

powered, a four-wheeled cart that could transport not only astronauts but also ample rock samples and lots of scientific equipment. The concept of a lunar rover originated with Wernher von Braun in the early 1950s. About a decade later, von Braun, now developing NASA's heavy-lift rockets with his team at Marshall, again wrote extensively about the need for a lunar roving vehicle.

Between 1965 and 1967 NASA engineers concluded that a vehicle of some type was critical to make the most of lunar studies from the Apollo program. By 1969 they chose the lunar roving vehicle as the winning concept and sanctioned its creation at Marshall, under von Braun's watchful eye. The final development and construction of the lunar rovers took place just before the Apollo 11 landing. Boeing oversaw the project, and General Motors, as a subcontractor, provided the wheels, electric motors, and suspension. The final cost of the lunar rover program, by the way, was $38 million, to provide four "cars." At $9.5 million each, price-wise, they definitely leave Ferraris and Lamborghinis in the dust!

This image of Mare Imbrium, one of the major lunar basins, was made by the Lunar Reconnaissance Orbiter years after the Apollo missions. It shows the basin as a large circular feature filling most of the frame. The triangle of prominent craters near the right edge of Mare Imbrium consists of Archimedes, Aristillus, and Autolycus. Just below and right of these craters lies Hadley Rille, landing site of Apollo 15.

The range of the lunar rovers was functionally limited to keep them in long-range walking distance of the LM in case a malfunction with the rover should occur. Nevertheless they really enhanced the functionality of the Apollo missions. These vehicles had a designed top speed of 13 kph (8 mph), but during Apollo 17's sojourn on the Moon, Gene Cernan reached a velocity of 18 kph (11.2 mph), making his drive the fastest on the surface of the Moon.

They weighed some 210 kilograms (460 pounds) and measured 3 meters (10 feet) from front to back. The frame was constructed from aluminum alloy tubing, and the three-part chassis was hinged so it could be folded inside the LM, to be unpacked and set up on the Moon's surface. The two side-by-side seats were made of aluminum tubing and had nylon webbing. An armrest separated the seats and each had footrests. The wheels consisted of a spun aluminum hub with tires made of steel strands woven together, coated with zinc, and covered with titanium chevrons to help achieve maximum traction on the powdery lunar surface.

Each wheel on the vehicle had its own electric drive, and the vehicle could be maneuvered by front and rear steering motors. Front and rear wheels could turn in opposite directions for maximum maneuverability, or they could be decoupled to provide specific aid in steering and traction. The energy to make this package move came from two silver-zinc potassium hydroxide batteries, each with 36 volts, that powered the drive and steering motors; a utility outlet was mounted on the front of the rover to power the communications equipment and a color TV camera. The whole thing could be controlled with a T-shaped handle by the astronaut-driver.

To deploy the rover, the astronauts had to use pulleys and braked reels, ropes, and cloth tapes. Stored in a bay inside the LM, the rover had to be released by an astronaut who climbed the LM's ladder, and then his companion could slowly unfold the chassis of the rover. The rover then moved down from the bay and could be set up on the ground. Rear wheels then could be locked in place. The entire frame could then come down and final assembly achieved. The astronauts then assembled the final aspects of the seats and footrests, switched on the vehicle's electronics, and the car was ready for a lunar test drive.

With the rover packed, along with everything else in the mighty Saturn V rocket, the crew of Apollo 15 prepared for launch on July 26. In mid-morning, the rocket shot skyward and the mission was underway. The launch did not go flawlessly, however. The first stage failed to shut off on time, introducing the possibility of the first stage engines colliding into the second stage boosters, which would have aborted the mission. Exhaust from the second stage engines also damaged an on-board telemetry package, but despite these initial concerns, the rocket performed adequately and lifted the spacecraft into an "Earth parking" orbit. After some two hours, the third stage ignited and sent the spacecraft shooting toward the Moon.

Following the three-day cruise to the Moon, the spacecraft passed behind the lunar far side and engines on the Command/Service Module burned for six minutes, placing the spacecraft into the right configuration for an LM landing at Hadley Rille. As they orbited the Moon, gazing down at its gray surface, they prepared for the undocking of the LM and its descent. On the first attempt to undock the Command/Service Module from the LM, however, the astronauts encountered a glitch. The Command/Service Module had a loose umbilical plug, it turned out, and it was reconnected by Worden. The separation then occurred, with Scott and Irwin in the LM and Worden remaining in the Command/Service Module.

The two descending astronauts stood in the tightly packed LM, as had their predecessors, for the slow downward movement. As they fell gradually toward Hadley

Rille, the astronauts discovered they were around six kilometers (3.75 miles) east of the target landing spot. Scott took manual control of the descent and maneuvered *Falcon* closer to the target. The craft gently dropped down just a few hundred meters from the intended site. Irwin announced that a landing probe on one of the legs touched the surface, establishing contact, and Scott cut the engine. This descent was faster than previous Apollo missions — faster than a hotel elevator — at approximately 2.1 meters per second (seven feet per second). "Okay, Houston. The *Falcon* is on the Plain at Hadley," declared Scott. One of the LM's legs landed in a small crater, a lunar landing first, tilting the craft by about ten degrees from level, so the LM appeared skewed in images, but this didn't create a functional problem. Missing the landing spot by a few hundred meters was also not alarming, since the rover made travel so much easier.

Rather than exiting the LM quickly, as in previous missions, Scott and Irwin stayed within *Falcon* for the rest of their landing day to stay on a normal sleep schedule. Before falling asleep, however, Scott photographed the lunar scene from the LM's top docking hatch, standing up and poking his head outside, and capturing the surroundings with a 500 mm lens. He recognized features the astronauts had studied on maps: the Mons Hadley Massif, Mons Hadley Delta (another massif), the Swan Range, the Silver Spur (a rock formation), Bennett Hill, Hill 305, and the North Complex, a group of hills. Scott returned through the hatch and repressurized the LM. And then Scott and Irwin went to sleep.

As the astronauts slept, Houston controllers became concerned with falling pressure readings from the descent stage oxygen tanks in the LM. They let Scott and Irwin sleep, however, but woke them an hour early to switch into a high-telemetry-rate mode. This allowed them to see that a valve was open, and they had lost a little less than ten percent of the oxygen. But the supply was still more than sufficient.

Once the astronauts were awake, they prepared for the first of three long moonwalks. Scott became the seventh human to climb down onto the Moon, and he said: "As I stand out here in the wonders of the unknown at Hadley, I sort of realize there's a fundamental truth to our nature. Man must explore. And this is exploration at its greatest." Some seven minutes later, Irwin joined him on the lunar surface, and the astronauts unpacked equipment stowed into the LM, including the rover. They positioned the TV camera, and then Scott became the first driver of a rover on

This view of the Apollo 15 Command/Service Module, taken on July 30, 1971, shows the craft in lunar orbit. It offers an opportunity to observe the open bay where panoramic and mapping cameras were located in the final three Apollo mission spacecraft. These cameras enabled sequential stereo coverage of the lunar surface, whereas prior missions captured stereo images only through the Hasselblad cameras operated by astronauts out of the Command or Lunar Module windows.

This perfect stereo view of the Lunar Roving Vehicle was made by Dave Scott during the third Apollo 15 moonwalk, at the Hadley Rille landing site. The western edge of Mount Hadley is visible at upper right, and the most distant lunar feature in this view is about 25 kilometers (16 miles) away.

the Moon's surface. He found that only the rear-wheel steering was working, and the suits didn't bend much when they sat down in the car, making driving a little strained. They reclined backward to ease the strain, which worked reasonably well.

Starting off at about 9 kph (6 mph), the astronauts began a planned circuit of activities. They drove along Hadley Rille, which made finding targets relatively easy. First, they arrived at Elbow Crater, a 340-meter (1,100-foot) depression. The mission now focused on geological activities, collecting samples, and photography. They continued on a distance of 500 meters (1,600 feet) to St. George Crater, a 2.4-kilometer (1.5-mile)

In this dramatic view, Apollo 15 astronaut Jim Irwin loads the lunar rover with tools and equipment as he prepares for the first lunar tour of the mission. The Lunar Module *Falcon* stands at left.

On August 1, 1971, Jim Irwin salutes the US flag on the Moon's surface during the second Apollo 15 moonwalk. The LM *Falcon* stands on the right-hand side of the image, and the view is oriented almost due south. In the background, Hadley Delta rises some 4,000 meters (13,000 feet) above the surrounding plain. The base of the mountain is about five kilometers (3.1 miles) away. The scene was photographed in 3-D by Dave Scott.

pockmark on the Moon. They did not see the amount of geologically intriguing ejecta they had hoped for, so instead of spending lots of time there, they moved on to a boulder they had noticed in the open, near a tiny crater. Taking samples, they tried to roll the rock and ended up chipping pieces off of it. Using a rake with tines set one centimeter apart they collected samples of pebbles from the area. They also took core samples by driving tubes down into the powdery lunar surface.

Moving again, Scott and Irwin bypassed a planned stop at Flow Crater because of time constraints and pressed on toward the LM. They stopped at Rhysling Crater and Scott spotted a large piece of vesicular basalt that he found irresistible. Although time was short, he stopped the rover, grabbed the rock, and carried it into the rover while Irwin distracted controllers in Houston. Because Scott pretended he had stopped to tighten his seatbelt, this rock was dubbed the "Seatbelt Basalt."

At the LM, the duo set up the normal ALSEP set of experiments with a seismometer, magnetometer, solar wind spectrometer, and other instruments. After more than 6½ hours of activity, the weary pair climbed back into the LM, ready to rest after such exertion. Irwin's water bag had not worked properly, and so he had gone through the whole surface activity without fluids.

Sixteen hours later, on August 1, Scott and Irwin commenced their second extravehicular activity. They would again focus on the Mount Hadley Delta, but this time headed directly to a site to the east of the previous day's action. They moved along the planned route but, unexcited by some of the visuals they encountered, skipped some of the planned sites. They took samples from a one meter (3.25 foot) diameter crater, finding breccias and porphyritic basalt. Scott then moved to a larger crater but found the rocks inside it too large to sample. Houston asked the pair to dig a trench to study soil characteristics, which Irwin did while Scott photographed the exercise.

They then returned to the rover and took a long drive to look at a three meter boulder, having noticed a greenish hue on it caused by magnesium oxide. They then traversed to Spur Crater, a 100-meter (320-foot) depression reaching a depth of 20 meters (66 feet) below the surrounding lunar surface. They took samples and noticed white mineral veins in some of them and a greenish hue in the soil that they surmised was caused by the gold color of their visors.

This unusual rock was collected by Dave Scott and Jim Irwin on the rim of Spur Crater, 50 meters (163 feet) above the surface of the mare on the slope of Mount Hadley Delta. They identified it as an anorthosite, a rock composed almost entirely of plagioclase feldspar, and believed it to be among the oldest of Moon rocks. Later dubbed "Genesis Rock," it turned out to be only about four billion years old, younger than the oldest Moon rocks, but the name stuck.

Then, in a moment of triumph, they collected what later became the most famous lunar sample taken from the Moon by any Apollo mission. What was labeled as Sample 15415 looked initially like a partially crystalline rock, but on further examination the astronauts saw it was nearly completely plagioclase, a tectosilicate feldspar mineral. At first they, and scientists who studied this sample, believed the explorers had found a piece of the lunar crust. The media dubbed the sample "Genesis Rock." Analysis later showed the rock to be 4.1 billion years old —about half a billion years younger than the Moon. Still, it remains a very ancient and fascinating sample.

Houston then asked the pair to collect as many small samples as they could. They

On August 12, 1971, Dave Scott gets a close look at the sample known as "Genesis Rock" at the Manned Spaceflight Center in Houston, Texas. At left, astronaut Joseph Allen looks on with interest. The whitish rock has been given permanent identification as specimen number 15415.

The "Fallen Astronaut" is an aluminum sculpture made by Paul Van Hoeydonck and left on the lunar surface by Dave Scott and Jim Irwin. It commemorates astronaut and cosmonaut colleagues who died in service: Theodore Freeman, Charles Bassett, Elliot See, Gus Grissom, Roger Chaffee, Ed White, Vladimir Komarov, Edward Givens, Clifton Williams, Yuri Gagarin, Pavel Belyayev, Georgy Dobrovolsky, Viktor Patsayev, and Vladislav Volkov.

raked for small samples, and Scott struck a rock with his rake, fracturing it, and he collected the resulting pieces. Before they climbed back into the LM, the two again tried to drill holes nearby to facilitate a heat-flow experiment, but had trouble as on the previous day. They set up the US flag near the LM and then returned to the interior after an excursion that lasted seven hours 12 minutes.

A third moonwalk — and drive — began with trying to rectify the problems the pair had with drilling holes. They extracted some core samples but continued to be frustrated with the drilling and didn't want to waste too much time focusing on it. They filmed the rover and then traveled to Hadley Rille, arriving at a small crater. Samples collected there were soft, and this area is thought to be the youngest explored by moonwalkers. They photographed the rille and looked for exposed bedrock, hoping to find ancient material. In the wall of the rille, they looked for layering, indicative of lava flowing over time through Palus Putredinus. Irwin found some exposed bedrock. He took samples. Scott found a coarse-grained basalt with large vugs, or cavities. He carted off a football-sized rock that came to be called "Great Scott," a 9.6 kilogram (21 pound) sample.

Back at the LM, with time running out, Scott had one more trick up his sleeve. "Well, in my left hand, I have a feather," he said. "In my right hand, a hammer. And I guess one of the reasons we got here today was because of a gentleman named

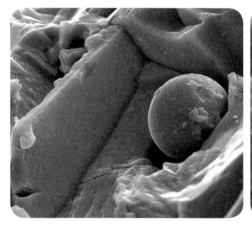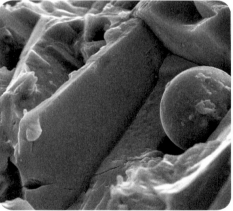

A Scanning Electron Microscope (SEM) stereo micrograph of the surface of a microbreccia – showing a glassy sphere and a platelet embedded into its surface. The sample was prepared from rocks returned to Earth by Apollo 11.

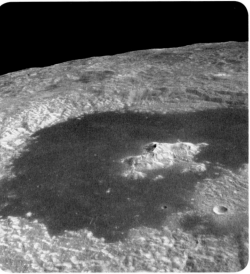

Crater Tsiolkovsky on the lunar far side is one of the most represented features in Apollo photography – we already saw Apollo 13's first stab at imaging it. This portrait was taken by Al Worden as the Apollo 15 Command Module orbited the Moon. The central peak stands out in stereo like a bright island in the dark sea of the crater floor.

Galileo, a long time ago, who made a rather significant discovery about falling objects in gravity fields. And we thought where would be a better place to confirm his findings than on the Moon. And so we thought we'd try it here for you. The feather happens to be, appropriately, a falcon feather for our *Falcon*. And I'll drop the two of them here and, hopefully, they'll hit the ground at the same time."

He dropped the two, and they hit the lunar surface simultaneously. A win for Galileo once again. Before ending the walk, Scott drove the rover away a distance and planted an object on the lunar surface. It was a sculpture, made of aluminum, created by the Belgian artist Paul Van Hoeydonck. Called Fallen Astronaut, it was a small astronaut figure, and was accompanied by the names of American and Soviet

On August 7, 1971, Dave Scott is helped out of the Apollo 15 spacecraft into a life raft by naval officer Fred Schmidt during recovery operations. The capsule landed some 530 kilometers (330 miles) north of Honolulu, Hawaii. One of the three parachutes failed to deploy properly, but the mishap caused no injuries.

explorers who died in the quest for space. After four hours 50 minutes, Scott and Irwin climbed back into the LM. Two days and 18 hours after they had landed, the astronauts blasted off from the Moon's surface, rejoined Worden, and headed back to Earth. The crew splashed down in the North Pacific on August 7, with 77 kilograms (170 pounds) of lunar samples on board. Apollo 15 had been the most scientifically interesting mission yet.

And with the Fallen Astronaut, the enmity between competing nations was well on the way to evolving into a special bond between brother explorers.

On the recovery ship USS *Okinawa* on August 7, 1971, the Apollo 15 crew, Commander Dave Scott (saluting), Al Worden, and Jim Irwin are happy to be back on Earth following their successful mission.

On April 16, 1972, the mighty Saturn V carrying the Apollo 16 astronauts launches from the Kennedy Space Center in Florida, bound for the Moon.

12

SERIOUS SCIENCE
(APOLLO 16)

With the success of Apollo 15, NASA could now set its sights on a follow-up mission that would again pursue broad scientific goals, but in a very different region of the Moon. The mission would be the most ambitious yet, aiming to land astronauts on the lunar surface for a full-on three days. It was scheduled for a launch in the spring of 1972, some eight months after the splashdown of Apollo 15.

The cultural tumult of the sixties by now seemed to be quieting down. The Vietnam War droned on, however, and peace talks in Paris were intermittent and characterized by only glimmers of hope. President Nixon made an historic visit to China, and, back home, signed legislation that would create a NASA effort to succeed Apollo: the Space Shuttle program. The US robotic spacecraft Mariner 9 relayed back pictures from Mars, and the Pioneer 10 spacecraft set off on a journey to the outer planets.

Apollo 16 aimed to improve our understanding of the Moon in a geological sense. Both Apollo 14 and 15 had visited areas that were associated with Mare Imbrium, one of the maria, or lunar "seas" – which at some relatively late stage in lunar history were filled in with lava. Scientists believed that studying geological processes in the older rocks would provide clues to the Moon's origin and early history. So Apollo 16 would set down in the so-called lunar highlands.

To best pursue this line of research, the scientists chose the Descartes Highlands, located west of Mare Nectaris. The Descartes region was intriguing because planetary scientists believed the magma that formed the region was viscous, and therefore older, than the smooth lava that later on filled the lunar seas. The long distance separating the region of Descartes from the other Apollo sites would be helpful in terms of spreading the instruments left behind for data collection.

The touchdown was planned to take place between two small craters, North Ray and South Ray, which spanned 1,000 meters (3,300 feet) and 680 meters (2,200 feet), and were believed to be relatively young. The exposed rock from these two small impacts would provide interesting material for the astronauts to examine. The so-called Descartes formation and the Cayley Formation were principal targets for study. Surprisingly to us now, many planetary scientists at the time believed these areas to be volcanic in nature. Impact geology was still a young science, growing in its sophistication.

The crew for Apollo 16 would be a blend of an experienced commander and two rookies. Commander John Young, age 41, was a San Francisco-born naval officer and aviator, test pilot, and aeronautical engineer. He had been selected as part of NASA's second astronaut group and served as pilot on Gemini III, Commander of Gemini X, and Command Module Pilot of Apollo 10, becoming the first person to fly solo around the Moon. He was Backup Commander for Apollo 13. His greatest moment came as Commander of Apollo 16, but Young's career would also stretch into the future, far beyond Apollo. He flew as Commander of two Space Shuttle missions in 1981 and 1983, thus achieving the longest career of any NASA astronaut. He was also the only person to have piloted and commanded four types of exploration spacecraft: Gemini, the Apollo Command/Service Module, the Apollo Lunar Module, and the Space Shuttle.

Prior to the mission, the Apollo 16 crewmembers train with their rover. Left to right: Lunar Module Pilot Charlie Duke, Commander John Young, and Command Module Pilot Ken Mattingly.

Young's deep expertise would be complemented by Command Module Pilot Ken Mattingly. Again healthy following his exposure to German measles ahead of the Apollo 13 flight, Mattingly now could look forward to his lunar mission at last.

Apollo 16's Lunar Module Pilot was Charlie Duke, at that time another rookie. Age 36, Duke was a product of Charlotte, North Carolina, and became an Air Force officer and test pilot. Selected within the group of NASA astronauts joining the fold in 1966, Duke was a member of the support ground crew for Apollo 10. On Apollo 11, he famously served as CAPCOM, exchanging pleasantries with Neil Armstrong and Buzz Aldrin while the world watched the first Moon landing. He served as backup Lunar Module Pilot on Apollo 13, but caught German measles before the flight, which exposed Mattingly to the disease. This shuffled the assignments and led to the pairing of Mattingly and Duke, along with Young, on Apollo 16. The mission's backup crew consisted of Commander Fred Haise, Command Module Pilot Stuart Roosa, and Lunar Module Pilot Edgar Mitchell. Call signs for the spacecraft were *Casper* for the Command/Service Module and *Orion* for the LM.

The launch of Apollo 16 was slated for March 17, 1972. However, for the first time in the history of the Apollo program, a technical problem caused a delay. Mission planners were concerned with the mechanism that separates the docking ring and the Command Module and also had concerns about the LM batteries. Moreover, during a test in January 1972, technicians damaged a fuel tank in the Command

Module. NASA employees rolled the rocket back to the enormous Vehicle Assembly Building, replaced the tank, allayed other concerns, and moved it back to the historic launch pad 39A. Now it was ready for launch, which was rescheduled for April 16.

Launch day dawned with a beautiful blue sky and just a smattering of clouds here and there. Just after noon, the mighty Saturn V blasted off and headed skyward. Just 12 minutes after liftoff, the spacecraft entered an Earth parking orbit. The crew checked the spacecraft carefully and prepared, as they orbited, for a third stage burn that would send them off toward the Moon. Small problems soon cropped up: the environmental control system did not seem to be functioning properly. The directional control of the third-stage burn also seemed a little off. But the crew overcame these challenges, lit the third-stage for a five-minute burn, and headed off toward a trans-lunar injection at some 35,000 kph (22,000 mph). Following the burn, the Command/Service Module, with the crew inside, maneuvered to retrieve the Lunar Module and then proceeded. However, the crew noticed that particles seemed to be coming off of the LM. Charlie Duke observed about five or ten particles detaching from a site where the "skin" of the LM appeared to be torn. The crew entered the LM and did not see any major issues from the inside. Once the cruise phase began in earnest, the crew placed the craft in "rotisserie" mode so that it would rotate three times per hour to spread solar heating evenly.

The crew could then rest. By the time flight controllers issued a wake-up call, the spacecraft was 181,000 kilometers (98,000 miles) from Earth, which would bring the spacecraft to the Moon in less than four days. Crewmembers conducted some experiments en route to the Moon. They studied electrophoresis, which analyzes the motions of dispersed particles in fluid under an electrical field. They collected data to see if particles

On February 9, 1972, the second rollout of the mighty Saturn V rocket with the Apollo 16 spacecraft attracted quite a crowd at the Vehicle Assembly Building at the Kennedy Space Center.

Charlie Duke relaxing during the delay while awaiting the second rollout of Apollo 16.

would move more effectively in zero gravity. They conducted a course-correction burn. They reentered the LM to again inspect it amid the strange particles they had seen coming off the craft. They did see additional paint, they believed, peeling off the LM. But the systems all seemed fine. Mattingly reported a "gimbal lock" warning light, which meant that the craft was not recording altitude data. Using the Sun and Moon as luminous targets, he realigned the guidance system, which fixed the problem.

As the third flight day began, the crew found themselves 291,000 kilometers (157,000 miles) from Earth, approaching the Moon, and with the spacecraft velocity decreasing. Crewmembers performed a so-called ALFMED test, the Apollo Light Flash Moving Emulsion Detector experiment, designed to determine why astronauts sometimes saw flashes in their eyes when the cabin was dark, even if their eyes were closed. The cause was believed to be cosmic ray strikes. They then suited up and practiced procedures for landing. As the third day ended, the craft began to speed up again, being pulled by the Moon's gravity.

On the fourth day, the crew readied for a lunar orbital insertion. Around 74 hours after liftoff, the craft passed behind the Moon and temporarily lost telemetry with Houston. The burn that occurred while the spacecraft was on the far side of the Moon placed it into an orbit that ranged from 108 to 316 kilometers (58 to 170 miles) from the lunar surface. The crew then readied for the maneuvers that would undock the LM and prepared for Young and Duke to descend for their exploration while Mattingly stayed aboard *Casper* and circled above them.

Waking on the fifth day, the crew encountered a problem, however. The retractable boom that was used to extend a mass spectrometer outward from the Command/Service Module was stuck in one position. Young and Duke would inspect the boom from the outside after their boarding of the LM, and so the crew prepared for the undocking. Slightly more than 96 hours into the mission, the undocking of the LM, *Orion*, from *Casper* took place. Mattingly readied to move the Command/Service

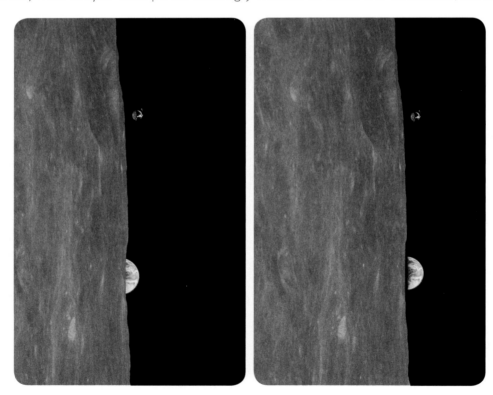

The Apollo 16 Command/Service Module floats off the lunar limb while Earth rises over the Moon in a striking stereo view. This was assembled from two sequential pictures taken from the Lunar Module *Orion* after undocking. The Command/Service Module's burn was delayed due to oscillations of the propulsion system, and *Orion's* landing was delayed until engineers determined the problem would not affect the return to Earth.

Module into a circular orbit while Young and Duke worked inside the LM to prepare it for its descent. Before the burn within *Casper*, however, an engine malfunction occurred. This called for a re-docking, but, after several hours of calculations and study in Houston, mission controllers decided to proceed with the landing. They were now six hours behind schedule and began the descent at the highest altitude of any Apollo LM — some 20 kilometers (11 miles). The engine burn took place, and by the time the craft was 4,000 meters (13,000 feet) from the target, Young could make out the landing site. Descending slowly, the LM touched down 270 meters (900 feet) north and 60 meters (200 feet) west of the planned landing site.

Once they touched down, Young and Duke took initial readings, removed their suits, powered down the LM, and gazed out onto the powdery gray lunar surface. The delay forced a compromise in the mission's schedule: the mission would reduce its lunar orbit time to less than a day following the surface activities, and the last of three moonwalks was cut down to five hours from the originally planned seven.

Following a period of sleep, Young and Duke prepared to begin their extravehicular activity. Young climbed out of the LM first, descending the ladder, and when he stepped onto the lunar surface, he exclaimed: "There you are: mysterious and unknown Descartes. Highland plains. Apollo 16 is gonna change your image. I'm sure glad they got ol' Brer Rabbit, here, back in the briar patch, where he belongs." Young thus became the ninth human being to walk on the surface of the Moon. Close behind, the 10th man, Charlie Duke, descended the ladder, the youngest person ever to walk on the Moon's surface. Duke was excited, exclaiming: "Fantastic! Oh, that first foot on the lunar surface is super, Tony!," his comment aimed at support crewmember Anthony England, in Houston.

Young and Duke first unloaded equipment, including the lunar rover and an ultraviolet camera and spectrograph. Young powered up the rover and discovered that the rear steering was ineffective. He set up the color TV camera, deployed the customary US flag, and the two astronauts prepared for exploration. They unpacked and set up the usual ALSEP experiments and instruments package, and while parking the rover, Young noticed that the rear steering suddenly began to work. After collecting rock samples in the immediate area, four hours into the moonwalk they piled into the rover and began to explore more distant areas.

Apollo 16 Lunar Module Pilot Charlie Duke salutes the US flag during the mission's first moonwalk on April 21, 1972. He is about to explore the Descartes Highlands; behind him stands the Lunar Module *Orion* and the lunar rover, with Stone Mountain in the background. John Young secured the images.

The first target of opportunity was Plum Crater, a small depression measuring 36 meters (120 feet) across, which was situated on the edge of a larger crater, Flag Crater. About 1.4 kilometers (0.9 miles) from the LM, they took samples from the crater's edge, scientists believing that this material may have emerged from relatively deep, from the Cayley Formation, far below the Moon's shallowest layer of regolith. Duke obtained and carted off, in the rover, the largest Moon rock taken by any Apollo mission, a brecciated rock nicknamed "Big Muley" after geologist William Muehlberger, Apollo 16 geology Principal Investigator. Cataloged as Lunar Sample 61016, this rock weighed 11.7 kilograms (26 pounds) and consisted of shocked anorthosite with a fragment of rock melted by impact. They moved on a relatively short distance to Buster Crater, where Duke photographed the region and Young established a magnetic field experiment. Young then demonstrated the maneuverability of the rover, which was filmed by Duke. They returned to the LM after more than seven hours of activity on the surface, reviewed their progress, and readied for more sleep.

The following day, Young and Duke prepared for their second moonwalk. They discussed the objectives with Houston, aiming to focus on a rise called Stone Mountain. They would climb in the area to reach a region called Cinco Craters, some

During the first Apollo 16 moonwalk, Charlie Duke collects lunar samples at the Descartes landing site in the lunar highlands. This image was shot by John Young as a portion of a panorama. Charlie is standing on the rim of Plum Crater, situated about 30 meters (98 feet) northwest of the *Orion* lander.

A closer crop of the scene shows the reflected image of John Young in Charlie's visor.

3.8 kilometers (2.4 miles) from *Orion*. This would place them at the highest elevation on the Moon attained by any Apollo astronauts, 152 meters (500 feet) above the lunar surface below. They set off on the drive/walk, traveling toward Stone Mountain, where the duo gathered rock samples. From the side of Stone Mountain, they had a spectacular view of South Ray Crater, a 700-meter-diameter (2,300 foot) blister. They spent nearly an hour in this region, photographing, gathering samples, and enjoying the incredible view.

Young and Duke then proceeded to the next target, Station 5, where they encountered a crater of about 20 meters (66 feet) in diameter. The objective here was to find uncontaminated, old material from Descartes that had not been mixed with younger ejecta from the impact that forged South Ray Crater. Geologists later stated that the samples collected at this site were probably from Descartes. In a similar fashion, the duo then traveled to Station 6, where they hoped to sample pristine material from the Cayley Formation. They continued to a low-elevation flank of Stone Mountain, where they spent an hour collecting samples. They next proceeded to Station 9, which had been nicknamed "Vacant Lot," an area geologists believed to be free of ejecta from the South Ray impact. Returning to the area of the LM, they stopped at the midway point

The lunar roving vehicle gets a high-speed workout by astronaut John Young in the "Grand Prix" run during the third Apollo 16 extravehicular activity. This view is a frame from a motion picture film exposed by a 16 mm Maurer camera held by Charlie Duke.

between the ALSEP experiment and home base. There they dug holes and conducted an experiment to test the strength and density of the lunar soil. Requesting a slight extension of time, Young and Duke got it and returned to the safety of the LM after seven hours 23 minutes, establishing the longest moonwalk to date. They repressurized the cabin, reviewed the work with Houston, had a bite to eat, and prepared for sleep.

The third moonwalk took place on the seventh day of the mission. The objective on this excursion was to visit North Ray Crater, the largest crater visited by Apollo astronauts, at 950 meters (3,000 feet) in diameter and with a depth of 240 meters (720 feet). Young and Duke had a smooth ride in the rover toward the crater, and they passed Palmetto Crater after a time, beyond which the strewn boulders became larger. When they arrived at the rim of North Ray Crater, they were an impressive 4.4 kilometers (2.7 miles) from *Orion*. This deeply dished crater was a fascinating sight for the visiting explorers.

They visited, studied, and photographed a large boulder they came to call "House Rock," which stood taller than four stories. They sampled the rock extensively, and these samples later provided one of the mission's primary surprises. When studied carefully, they disproved the theory of an origin in volcanism, showing a clear history of impacts. Countless small holes in the rock showed a long history of micrometeoroids striking the rock for a very long period.

After 1½ hours at the site, Young and Duke moved on to Station 13, a boulder field near the site of North Ray Crater. They stopped at an intriguing boulder, which stood 30 meters (100 feet) high, called "Shadow Rock," where they sampled soil that appeared to be in a spot permanently shadowed from sunlight. During this time,

In this spectacular still image from the color TV transmission, the Apollo 16 Lunar Module *Orion* lifts off from the Descartes Highlands on April 22, 1972. Following this departure, only one more Moon mission was on the books.

Young drove the rover to a lunar speed record of 17.1 kph (10.6 mph), as the rover sped downhill. They returned to the LM three hours later and readied for the return to see Mattingly. Before leaving, Duke placed a picture of his family on the Moon's surface, not far from the LM. At Station 14, Cat Crater was named for Duke's sons, Charles and Tom. And at Station 16, Dot Crater was named for his wife, Dorothy.

On April 24, the ascent stage of the LM lit and Young and Duke lifted off toward Mattingly. The rendezvous and re-docking went without a hitch, and the three were on their way back to Earth. During the return cruise, Mattingly performed a spacewalk to retrieve film cassettes from the exterior of the Command/Service Module. The crew splashed down in the Pacific Ocean, southeast of Christmas Island, on April 27, and was recovered by the USS *Ticonderoga* after more than 11 days in space.

Apollo 16 had achieved its goals and pushed the investigation of lunar geology farther than ever before. Back on Earth, studies of the rocks they had gathered put paid to the idea of lunar volcanism as a major force in shaping the Moon's surface and gave rise to a surprising new theory of the origin of the Moon.

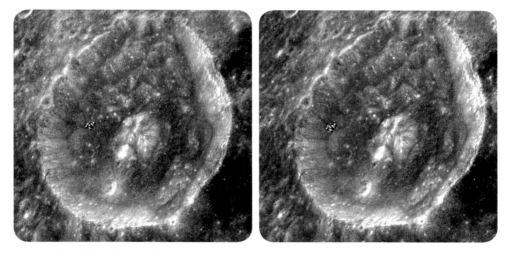

The tiny flying object in the foreground of this stereo image, seen toward the left side of the crater Schubert B behind it, is *Orion*, the Apollo 16 Lunar Module ascent stage. Astronauts John Young and Charlie Duke are aboard, returning from the lunar surface. Ken Mattingly shot these pictures sequentially from the orbiting Command Module, *Casper*, and *Orion* shifted just enough laterally to pop out in stereo.

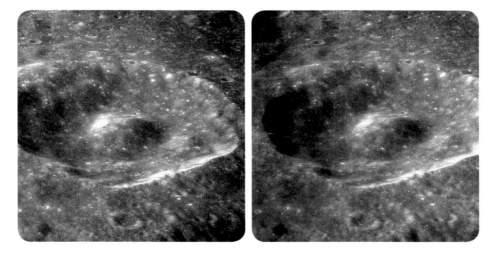

With an unusual central peak, the crater Alpetragius spans 40 kilometers (25 miles). Perplexed planetary scientist Gerard Kuiper, who called it "an egg in a bird's nest," hypothesized it might be covered with ejecta from a nearby impact. The Apollo 16 astronauts photographed the crater from lunar orbit.

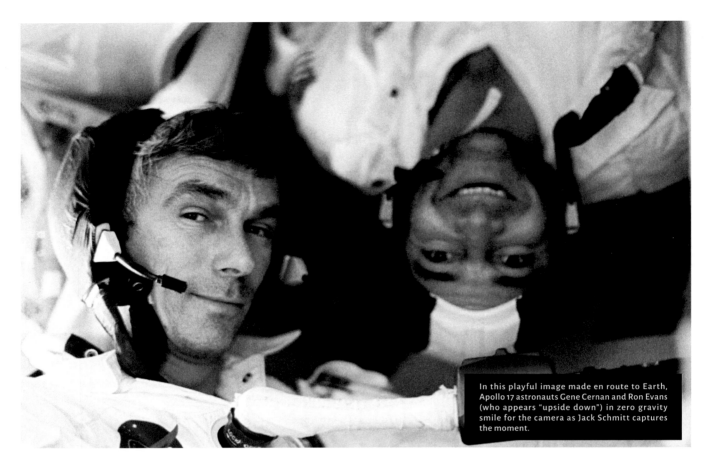

In this playful image made en route to Earth, Apollo 17 astronauts Gene Cernan and Ron Evans (who appears "upside down") in zero gravity smile for the camera as Jack Schmitt captures the moment.

On December 13, 1972, Jack Schmitt smiles inside the LM, *Challenger*, following the third lunar moonwalk. He was photographed by Gene Cernan.

13 ARRIVEDERCI LUNA
(APOLLO 17)

Following the enormous scientific success of Apollo 16, NASA planned Apollo 17 in late 1972 as another so-called J-type mission, with three days on the lunar surface to achieve more lofty science goals. The target lunar area would be chosen to complement what the previous two missions had accomplished. Astronauts this time headed for the Taurus-Littrow Valley, a depression along the southeastern edge of Mare Serenitatis. A ring of mountains in this area was believed to have formed when a very large impact created the "sea," and the valley would give astronauts the opportunity to sample material that was very old from highlands surrounding the landing site. They also planned to investigate what NASA believed was relatively recent volcanic activity in the region.

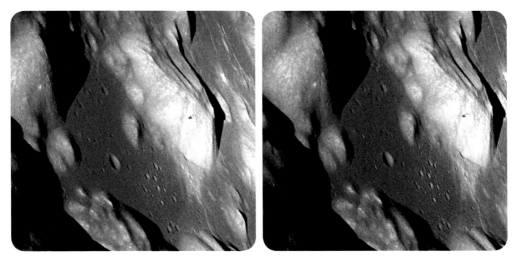

This spectacular view of the Taurus-Littrow Valley landing site for Apollo 17 was taken by Commander Gene Cernan from the LM, *Challenger*, during the last orbit before descending to the lunar surface. The Command/Service Module, *America*, can be seen on the right, flying above the South Massif. This view encompasses the entire area the astronauts explored, including Camelot Crater and the Lee-Lincoln Scarp.

Apollo 17 marked a number of important milestones for NASA. This was the first mission, with its increasing interest in sending scientists to the Moon, to include no test pilots. It was to set a record as the longest duration Moon landing, the longest aggregate time of moonwalks, the longest time in orbit about the Moon for any mission, and returning the largest amount of lunar samples to Earth. It was also, of course, the last time, to date, that humans have walked on the Moon. More on that later.

The Apollo 17 crew had a dynamic composition. It consisted of Commander Gene Cernan, Command Module Pilot Ron Evans, and Lunar Module Pilot Harrison "Jack" Schmitt. Again, NASA had paired an experienced spaceflight veteran, Cernan, with two rookies. Cernan, now aged 38, had been selected as one of the third group of NASA astronauts in 1963. Originally selected with Tom Stafford as the backup crew for Gemini IX, Cernan was thrust forward into that mission when the primary crewmembers, Elliot See and Charles Bassett, were killed in the tragic plane crash in St. Louis in 1966. Further, he was selected as backup Lunar Module Pilot for Apollo 7, even though that mission carried no LM. That selection, however, placed him into position as Lunar Module Pilot for Apollo 10, and in 1969 he participated in that mission as the final dress rehearsal before the Moon landing. He maneuvered the LM, *Snoopy*, to within 16 kilometers (8.5 miles) of the Moon's surface, gaining valuable experience as a spacecraft pilot.

A native of Kansas, Ron Evans, aged 39, was a naval officer and aviator, electrical engineer, and aeronautical engineer. He completed flight training through the Reserve Officers' Training Corps (ROTC) program at the University of Kansas, and in the early 1960s experienced a great many flight hours on duty in the Pacific. A Vietnam War veteran, he was selected as a member of NASA's 1966 astronaut group. Evans served in support groups for the Apollo 7 and Apollo 11 missions and was selected to be backup Command Module Pilot for Apollo 14, which placed him in the rotation to be selected for the Apollo 17 prime crew.

Age 37, Jack Schmitt was born in New Mexico and became a geologist. He accomplished his undergraduate work at Caltech before receiving a PhD in geology from Harvard University in 1964, having studied formations in Norway. Selected in the group of scientist astronauts in 1965, he was at that time an experienced veteran of the US Geological Survey's Flagstaff branch in Arizona, where the great scientist Gene Shoemaker was training prospective Apollo astronauts around places such as Meteor Crater. Following his astronaut selection, he spent a year in Air Force training learning how to fly a jet. He returned to Mission Control in Houston and helped train other Apollo astronauts in the vagaries of geological features. He focused on helping to train the astronauts to observe geological features from lunar orbit, as well as to make the most of their specimen collections on the surface. Schmitt was subsequently selected as part of the Apollo 15 backup crew.

Schmitt's career was tied to the overall situation with Apollo's future. NASA had originally planned more Apollo flights than actually occurred. The agency hoped to fly missions to the Moon through Apollo 20, which would complete a robust scientific exploration of our nearest celestial neighbor, they believed. However, concerns began to creep through Congress as early as Apollo 1's devastating fire. This tragedy forced a reboot of the program. The agency then planned its developmental missions that led to the triumphant landing of Apollo 11. Following that milestone, NASA had plans for nine additional Moon landings, matching the number of Saturn V rockets they had remaining. The four missions that followed Apollo 11 were planned to be the standard mission types, through Apollo 15. Five more missions, through Apollo 20, were planned as the extended J-class missions, with the three-day stay on the lunar surface, the lunar rover capability, and a more robust scientific program. Originally, Apollo 16 was planned to focus on the Descartes Highlands, Apollo 17 the Marius Hills, Apollo 18 Copernicus Crater, Apollo 19 Hadley Rille, and Apollo 20 Tycho Crater. This would complete NASA's vision for lunar exploration.

But budget cuts intervened and plans changed. During the first days of 1970, NASA administrators initially announced the cancelation of Apollo 20 so that its rocket could be used to launch a newer pet program, the Skylab Space Station, and then in the fall of that year the cancelation of Apollos 18 and 19 was announced as well.

Back to Jack Schmitt. The standard rotation placed him in the active crew for what was planned as Apollo 18. When this mission was canceled, Schmitt was moved up to Apollo 17 largely due to pressure from the scientific community, demanding that a geologist fly on the one remaining Apollo journey. So astronaut Joe Engle, originally slated for the mission, was unseated, and Jack Schmitt was bound for the Moon.

Now, with everyone aware that Apollo 17 would be the final lunar mission, the landing site choice became super-critical. NASA planners reconsidered all possibilities. They thought about the crater Copernicus but believed in the end it could be bypassed because samples from Apollo 12 would reflect the geology of that impact. They considered landing in the lunar highlands near another famous crater, Tycho, but were worried about the roughness of the terrain there and the possibility of a difficult or dangerous landing. They were also fascinated with the far-side crater Tsiolkovsky, named for the great Russian rocketry pioneer, but were worried about the complexity of the landing there and also about the difficulty of communications from the lunar far side. They also considered landing in the area of Mare Crisium but were reluctant to operate in an area that could be visited by Soviet craft. Luna 21, with its rover Lunokhod 2, did land in the area in 1973.

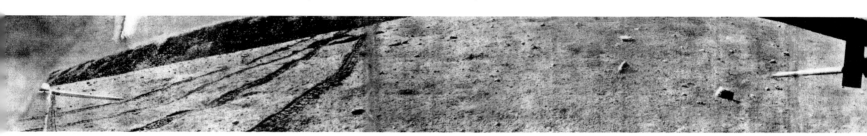

Lunokhod 2's tracks in the Mare Crisium following a successful mission by Luna 21 in 1973.

The final set of choices came down to three other sites entirely. NASA planners were interested in the craters Alphonsus and Gassendi, the first being an ancient scar and the second a crater nearly completely filled with lava, and the Taurus-Littrow Valley. The latter offered an opportunity to collect samples of ancient material from the lunar highlands, far away from Mare Imbrium, the area of previous visits, and to sample volcanic basalts from a relatively young lunar epoch. Taurus-Littrow, enabling both of these objectives, in the end became the target of choice.

Flight Director Gene Kranz at Mission Control, Houston.

On December 6, 1972, the last Saturn V rocket to see action in the Apollo program, stood on the historic pad 39A at Kennedy, ready to blast off. The mission was scheduled for an evening launch, but the countdown was held at T-minus 30 seconds due to an automatic cutoff in the launch sequence. Flight controllers identified a minor problem, and the launch countdown was reset and restarted. Finally, at 12:33 a.m. on December 7, the rocket ignited and carried the three astronauts skyward in what was the first and only nighttime launch of an Apollo mission. The night liftoff was dramatic; half a million people witnessed it in the area of the space center, and the streak from the rocket's trail was visible widely across Florida.

Following launch, the astronauts fired the third stage around three hours into the mission, which set the spacecraft on course toward the Moon. The component call signs were again typical of NASA administration: *America* for the Command/Service Module and *Challenger* for the LM. After the customary three-day cruise, Apollo 17 approached our celestial neighbor, and the crew fired the service propulsion system engine in the Command Module to slow the craft down and ready it for a lunar orbit insertion. In orbit, the astronauts examined the lunar surface below and prepared for the LM separation and descent.

Cernan and Schmitt climbed into the LM and prepared to set their sights on the Taurus-Littrow Valley, while Evans remained in the Command/Service Module. Evans would make routine observations, keep an eye on his comrades below, and perform some experiments during the three days of separation from his friends. Cernan and Schmitt stood in the crowded LM, began their powered descent, moving slowly downward to the lunar surface, and pitched *Challenger* over to gain their closing first view of the Taurus-Littrow Valley below. Cernan piloted the craft while Schmitt navigated, monitoring data on the craft that was needed to position the LM correctly for a soft touchdown. On December 11, in the middle of the afternoon, Eastern time, the LM made contact, and the last lunar explorers were on the Moon. There would be three moonwalks during Apollo 17, setting up and running an expanded suite of experiments, of course collecting more lunar samples and capturing images of the area.

Cernan and Schmitt exited the LM, configured the craft for their activities, and began to establish the experiments. They unpacked and assembled the last of the lunar rovers. They checked and set up various pieces of equipment. This mission

Gene Cernan with the parked lunar rover. This pair was created with pictures taken by Jack Schmitt during the third Apollo 17 moonwalk. The bright pattern on the high-gain antenna is sunlight reflected off the top of the TV camera, which was covered by mirror tiles and equipped with a sunshade attached to the top of the lens.

On December 12, 1972, at the Taurus-Littrow Valley landing site, Gene Cernan salutes the largest US flag deployed during the Apollo missions. Behind him stands the LM, *Challenger*, and the lunar rover that traversed the greatest distance of any on the Moon's surface, some 35 kilometers (22 miles). This stereo pair was taken by Jack Schmitt, the only geologist to walk on the Moon. The long baseline between the two camera positions for these images produces a miniaturizing effect that makes the scene look rather like a LEGO project!

On December 14, 1972, the astronauts parked the lunar rover at its final resting place, about 145 meters (476 feet) east of the LM, *Challenger*, visible behind on the left. The spot was carefully chosen so its TV camera, controlled from Houston via a large high-gain antenna, could film the last Apollo ascent from the lunar surface and broadcast it live to viewers on Earth. You might notice a little "bead" between the rover's antenna and the LM: that is Jack Schmitt, in the distance, who has moved between the two shots, making him appear to float in space right in front of us.

On December 12, 1972, in a view taken into the Sun, the rover sits parked on the eastern rim boulder field of Camelot Crater, about one kilometer (0.6 miles) west of the Apollo 17 landing site. The crater is 650 meters (2,130 feet) across. The Sculptured Hills lie in the distance. This stereo view was made "accidentally" by Gene Cernan as part of a set of photos for a panorama.

brought more experiments than any predecessor. There was the Traverse Gravimeter, a device that would be carried on the rover, and would record gravitational fields, providing potential clues about the Moon's internal structure.

There was also the Scientific Instrument Module. This package contained three experiments: a lunar sounder, an infrared scanning radiometer, and a far-ultraviolet spectrometer. The sounder was used for creating a geological map of the Moon in the area of the mission, to a shallow depth, by sending electromagnetic pulses into the surface. This would create a map of sorts to a depth of about 1.3 kilometers (0.8 miles). The radiometer had the objective of creating a temperature map of the Moon's surface to help sort out various geological features such as densities in the crust, fields of dense rocks, and so forth. The astronauts would use the spectrometer to study the Moon's atmosphere, capturing data on its density and composition. The instrument package also contained a panoramic camera, a mapping camera, and a laser altimeter.

More instruments abounded. The investigation that had started on Apollo 16 into the flashes of light Apollo crews noticed in a dark capsule with their eyes closed continued. They averaged about two per minute and were notable both during the cruise phases to the Moon and back and in lunar orbit. Cosmic rays were confirmed as the source of this experience.

The astronauts also brought along a biological cosmic ray experiment nicknamed BIOCORE that consisted of five mice with implanted radiation monitors. Four of the five mice survived the mission, and the study later found lesions on the scalps and livers of these small mammals. They were not believed to have been caused by cosmic rays, and autopsies of the mice found no apparent affects from cosmic ray exposure. Cernan dubbed the mice "Fe," "Fi," "Fo," "Fum," and "Phooey."

Yet another experiment package contained the so-called Surface Electrical Properties Experiment (SEP), which included two parts, an antenna for transmission and a receiving antenna, widely separated, the latter placed on the rover. Electrical properties of the regolith (lunar soil) could be studied by comparing the transmitted and received signals. This experiment helped to show that the shallowest level of powdery lunar surface, extending down about 2 kilometers (1.2 miles) deep, is remarkably dry.

In the evening of December 11, Cernan and Schmitt prepared to commence their first moonwalk. Four hours after they touched down, Cernan and Schmitt climbed down onto the lunar surface, becoming the eleventh and twelfth persons to walk

The Apollo 17 rover's repaired fender. The repair was necessary to keep the excessive quantities of dust thrown up by the rover's wheels away from the astronauts and their equipment.

on the Moon, and the last to this day. The pair offloaded the rover, setting it up, and Cernan accidentally broke off part of the fender by brushing against it with a rock hammer, tearing off part of the fender's extension. This caused the pair to be covered with a fair coating of Moon dust when the rover was traveling, and a similar incident had taken place on the previous mission, Apollo 16. They attempted to fix the problem by using duct tape to reattach the fender extension, but lunar dust stuck to the tape, diminishing its adhesive properties. Following the first moonwalk (drive), the astronauts consulted Houston and the team came up with a better attachment of the tape, which helped.

Cernan and Schmitt then deployed the standard ALSEP experiment package in close proximity to the LM and set off on their lunar drive. They moved south of the landing site to Steno-Apollo, a 520-meter (1,700-foot) diameter pockmark on the lunar surface. There, they began to collect rock samples, capturing about 14 kilograms (31 pounds) during this drive, and they set up two experiment packages that contained explosives. These were detonated to test geophones, which measure ground movement, and seismometers from previous missions. They also made measurements associated with the gravimeter. After more than seven hours, Cernan and Schmitt returned to the LM to rest and record what they had accomplished.

The following day, the two commenced their second moonwalk. This, too, lasted more than seven hours. They first repaired the long-suffering right rear fender on the rover, taping maps together in an assembly with duct tape and clamping the new part onto the fender, which helped to prevent lunar dust from coating them as they drove from place to place. This time they accumulated samples from a variety of geological forms and places. They focused on so-called avalanche material at the base of South Massif, and were attracted by orange-colored soil at Shorty Crater, a probable volcanic feature stretching 110 meters (120 yards) across and some 14 meters (15 yards) deep. Basaltic flows including what appeared to be orange-colored, glass-like deposits intrigued the astronauts here. They also explored Camelot Crater, which afforded ample ejecta for sampling. They brought 34 kilograms (75 pounds) of samples back to the LM during this excursion.

The following day, Cernan and Schmitt commenced the last moonwalk of the Apollo era. They collected more samples, made more gravimeter readings, and explored a continuum of interesting sites: the North Massif, Van Serg Crater, which spanned

On December 12, 1972, during the Apollo 17 second moonwalk, Gene Cernan took a series of photos to be assembled into a panorama he titled "Geology Station 5 at Camelot Crater." At that moment, Jack Schmitt was heading for the lunar rover, carrying the sample scoop in his left hand. The East Massif is on the left and Bear Mountain on the right. Schmitt was present in only one frame and thus appeared as a "ghost" in this stereo, but careful image processing restored him to solid form!

Boulders stand in the foreground of this view made on the southern rim of Camelot Crater on December 12, 1972. The North Massif lies in the background. The rocks were ejected by the impact that formed Camelot Crater, and their angular shapes and relative lack of dust coating are evidence of the young age of the crater, about 70 to 80 million years. This stereo view was again assembled from the images made by Gene Cernan for his panorama at Geology Station 5.

Rocks and boulders lie strewn across the western rim of Camelot Crater in this stereo view. The crater was named by the astronauts after the legendary castle of King Arthur. Some of these rocks show bluish hues; on the Moon these are generally due to titanium-rich minerals. The Lee-Lincoln Scarp can be seen in the distance, traversing the Taurus-Littrow Valley from left to right, and to the side of the North Massif, near the center of this view.

100 meters (325 feet), and explored the so-called Sculptured Hills. The astronauts collected a brecciated rock as a dedication to the several nations working together for the exploration of space and unveiled a spacecraft plaque commemorating Apollo's achievements.

Jack Schmitt then climbed back into the LM. Following him, Gene Cernan became the last human being to set foot on the Moon, at least as of this book's publication. That span of 46 years is incomprehensible to many – half a century without further *in situ* exploration of the Moon's surface. Before he climbed up the LM's ladder, Cernan said: "I'm on the surface, and as I take man's last step from the surface, back home for some time to come — but we believe not too long into the future — I'd like to just [say] what I believe history will record. That America's challenge of today

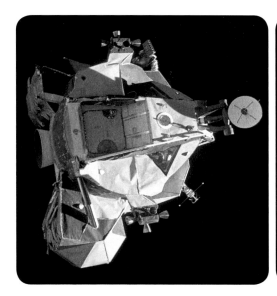
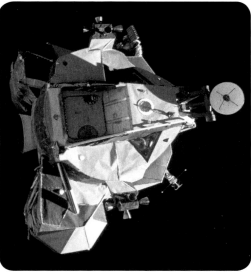

As the Apollo era comes to a close, the ascent stage of the LM, *Challenger*, carries Gene Cernan and Jack Schmitt upward, back to rendezvous with the Command/Service Module, from which Ron Evans made these pictures. The view was taken moments before docking the two craft. Looking fragile and irregularly shaped, the LM had thin walls in order to save weight and was incapable of flying through Earth's atmosphere. Note the ascent rocket engine on the left, the round rendezvous radar antenna on the right, and the square hatch through which the last lunar adventurers climbed back to start their journey home.

has forged man's destiny of tomorrow. And, as we leave the Moon at Taurus-Littrow, we leave as we came, and God willing, as we shall return, with peace and hope for all mankind. Godspeed the crew of Apollo 17." Then Cernan climbed into the LM.

On December 14, Cernan and Schmitt fired the LM ascent engine and left the lunar surface. They rejoined Ron Evans, docking with the Command/Service Module in lunar orbit, and the crew stored all materials to be returned within *America*. They jettisoned the LM and prepared to fire the engine to carry them back to Earth. At the end of the homeward cruise, the Command Module landed in the Pacific Ocean, splashing down triumphantly, with a recovery by USS *Ticonderoga*. On December 19, 1972, the astronauts began their re-assimilation to Earth and delivered their valuable finds to a grateful nation.

That was it. The Apollo program was finished. At least for half a century, humans had traveled to the Moon for the last time.

The Zodiacal Light (ZL) photographed from Apollo 17. In common with the three previous missions, Apollo 17 passed through the "double umbra" – on the far side of both the Moon and the Earth – the darkest place mankind has yet visited. Just before sunrise, it was possible to photograph the Sun's inner family and "atmosphere." We were able to make this stereo from two exposures about two days apart. In this time, the planet Jupiter (top left) moved against the background stars, so it appears to stand out. The cone-shaped appearance of the ZL is due to reflected sunlight from millions of dust particles orbiting roughly in the plane of the solar system. The motions of these grains were the subject of Brian May's Imperial College PhD studies, on hold at this time due to the interference of rock music, but completed 30 years later.

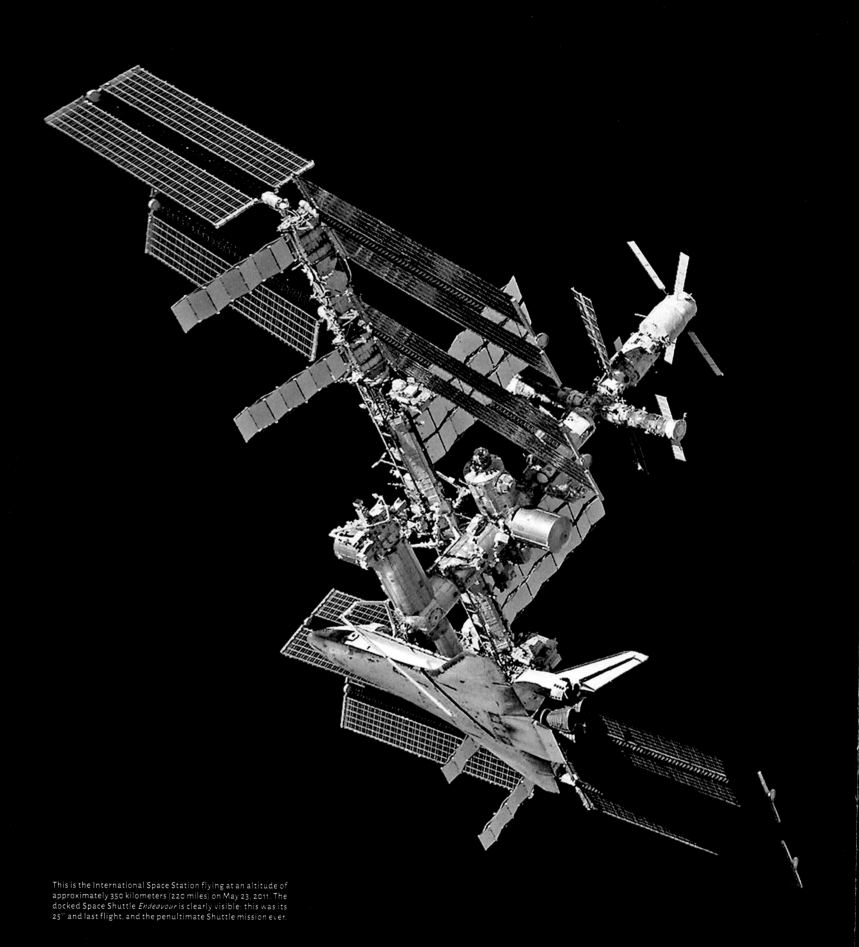

This is the International Space Station flying at an altitude of approximately 350 kilometers (220 miles) on May 23, 2011. The docked Space Shuttle *Endeavour* is clearly visible: this was its 25th and last flight, and the penultimate Shuttle mission ever.

14 AFTER THE RACE

Even as Apollo wound down, Apollo–Soyuz wound up. The Apollo–Soyuz Test Project, a cooperative venture between the Soviet Union and the United States, took place in 1975. The first joint spaceflight between the two superpowers, the mission lasted five days (Soyuz) and nine days (Apollo), and brought together the two spacecraft in orbit and a cast of explorers: Commander Tom Stafford, Command Module Pilot Vance Brand, and Docking Module Pilot Deke Slayton, together with Commander Alexei Leonov and Flight Engineer Nikolay Rukavishnikov.

Astronaut Deke Slayton embraces cosmonaut Alexei Leonov in the Soyuz spacecraft on June 17, 1975 when the two cold-war rivals' respective spacecraft rendezvoused and docked, and a new era of cooperation in space began.

The two craft launched within less than eight hours of each other, docking two days later. Three hours later, Stafford and Leonov took part in the first international handshake in space, which took place while the spacecraft were orbiting high over the French landscape. The crews read a statement from Soviet Premier Leonid Brezhnev and talked with US President Gerald Ford — Nixon had finally been ousted for his Watergate crimes the year before. The five men dined together, exchanged flags and gifts, made numerous jokes, and explored the ships. Because of Stafford's pronounced accent, Leonov later said the men had spoken three languages, Russian, English, and "Oklahomski."

The Hubble Space Telescope as seen in orbit by the Space Shuttle Atlantis's crew on May 19, 2009, when it was released for the final time after being serviced. Launched in April 1990, Hubble is to date the only telescope designed to be repaired and upgraded by astronauts in space, with five servicing missions flown by NASA Space Shuttles between 1993 and 2009. Hubble is still fully functional as of 2018, and engineers believe it could remain operational well past the launch of the James Webb Space Telescope in 2020, hopefully allowing joint observations and opportunities for spectacular 3-D imaging!

The Space Shuttle *Discovery* "poses" for a thorough series of inspection
photos before approaching the International Space Station for docking.

Space Shuttle *Atlantis* and its six-member STS-132 crew head toward Earth orbit and rendezvous with the International
Space Station. Liftoff was at 2:20 p.m. (EDT) on May 14, 2010, from launch pad 39A at NASA's Kennedy Space Center.

The exercise confirmed the rapidly increasing cooperation between the nations,
although for the Americans it would prove to be the last manned spaceflight until
the Space Shuttle program began six years later. Russians and Americans would
spend plenty of time together years later, in the International Space Station (ISS).

THE ERA OF THE SPACE STATION

The past generation of space exploration has been characterized by vehicles with
destinations other than the Moon. The American space shuttles dominated the scene
between the 1980s and their last flight in 2011. The genesis of the shuttle program
actually extended back to 1969, even before the Apollo 11 Moon landing. The first
shuttle launch in 1981 opened a new frontier in exploring low-Earth orbit, working
scientific problems in zero gravity, monitoring Earth as a planet, and conducting
numerous scientific tests and experiments.

The core module of the Soviet space station Mir was launched in February 1986, and the station kept growing with the addition of more modules for ten years. It was eventually de-orbited and destroyed in March 2001. American astronauts spent nearly 1,000 days living and working in orbit with Russian cosmonauts, and space shuttles rendezvoused ten times with Mir. The Shuttle–Mir program paved the way to the International Space Station and began an era of cooperation and exploration rarely seen in human history.

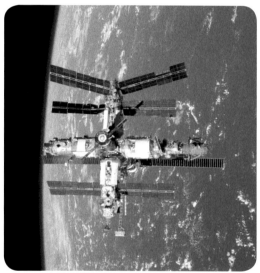

This is Russia's Mir Space Station, backdropped by Earth's horizon, as Space Shuttle *Endeavour* performed its fly around following undocking on January 29, 1998. This stereo view was assembled from two pictures documenting NASA's mission STS-89, the eighth of nine missions to dock an American Space Shuttle with the Russian Mir Space Station.

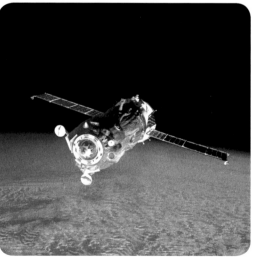

Soyuz TM-26 had just undocked from the space station Mir when the pictures this stereo view was derived from were taken, on February 19, 1998. Cosmonauts Anatoly Solovyev and Pavel Vinogradov were on board, together with French astronaut Léopold Eyharts. Solovyev and Vinogradov had just accomplished their mission to perform emergency repairs to save Mir after the collision with a Progress cargo spacecraft.

The Soviets conducted their own highly successful new branch of space exploration in low-Earth orbit with the launch of the Mir Space Station in 1986. This ambitious program constituted the first modular space station ever constructed and was built in orbit between 1986 and 1996. Six modules were ultimately added to the core that had been launched originally. As with the American shuttle program, Mir gave the Soviets, and after the fall of the Soviet Union the Russians, a platform for conducting science, testing materials, monitoring Earth science, and enabling zero gravity to help further scientific experiments. It was the first continuously inhabited research facility in space and held the record for the longest continuous human presence in space, at 3,644 days, until its successor captured a new record in 2010.

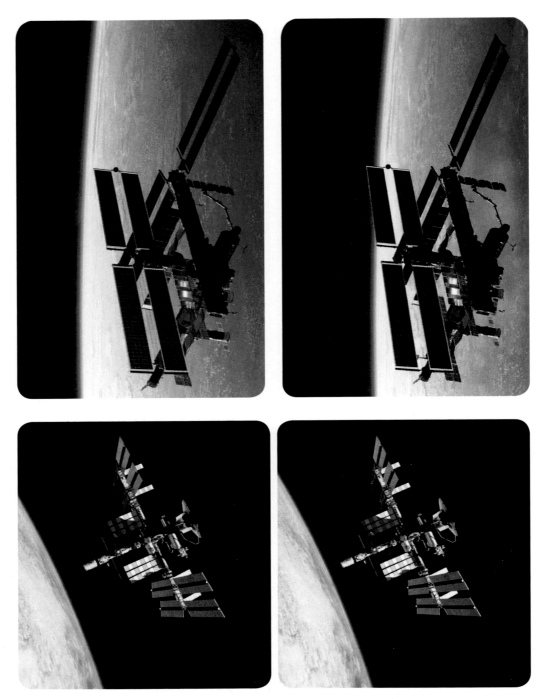

Step by step since 1998 the ISS has grown into the largest spacecraft ever built. These stereo views show the ISS in September 2006, after the STS-115 assembly mission of Space Shuttle *Atlantis*, and in May 2011 with the docked Space Shuttle *Endeavour* on top of it. These 3-D pictures have been assembled from images taken by crews leaving the station.

As Mir was withdrawn, plans for the International Space Station (ISS), a next step in floating a permanent, Earth-orbiting laboratory, took hold. A joint project between NASA, Roscosmos (Russia), JAXA (Japan), the European Space Agency (ESA), and the Canadian Space Agency (CSA), the first component of the ISS was launched into orbit in 1998. The ISS continues to operate and the cooperating space agencies expect it will last until about 2028. Since the arrival of a crew in 2000, the ISS has been continuously occupied, for more than 17 years and counting. Some highlights of the ISS's research and exploration include scientific research in a wide variety of areas, such as medicine in zero gravity, life sciences and astrobiology experiments, observations of Earth and its changing climate, and materials science experiments that have improved daily life on our planet.

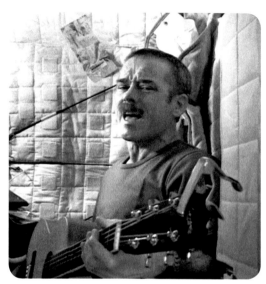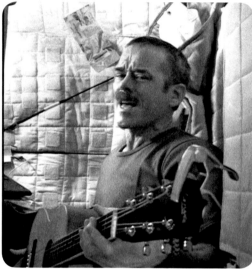

Commander Chris Hadfield performs his version of David Bowie's "Space Oddity" on board the ISS. Filmed to mark the reentry from his final space mission and posted on May 12, 2013, it was the first music video made in space, watched by 40 million people. David Bowie himself described Chris's cover of Space Oddity as "possibly the most poignant version of the song ever created."

The ISS has marked significant moments with cultural highlights, too. The enormously popular and accomplished Canadian astronaut Chris Hadfield created one of these special moments when, in 2013, he produced a music video in orbit as he sang and played the great David Bowie song "Space Oddity." For many, this moment seemed to crystallize a cultural feeling of what the space program, reaching for a greater understanding of the universe, is all about. Interestingly, many years before, David Bowie had recorded "Space Oddity" at Trident Studios in London in 1969. The song predated the Apollo 11 Moon landing. And, astonishingly, the day after Bowie recorded the song, in the very same studio, Brian May and his band Smile, the precursor to Queen, recorded the song "Earth" in the very same spot — and both songs, by coincidence, discussed space travelers!

Another, more recent astronaut has helped to lift public consciousness of the ISS and its activities. British space traveler Tim Peake has become another tireless spokesperson for space exploration, writing about his amazing experiences in books, delivering unique talks, and helping to share what it's like to stare down on this fragile blue planet.

Celebrating life through music has now happened in space. Years ago, when Apollo was in full gear, the pinnacle of music festivals was Woodstock, which introduced so many important acts to the world in that huge festival setting in New York in 1970. Some years later, however, music and culture came together again in what was arguably an even more successful festival, and one that helped humanity.

Tim Peak and Brian May, June 2018.

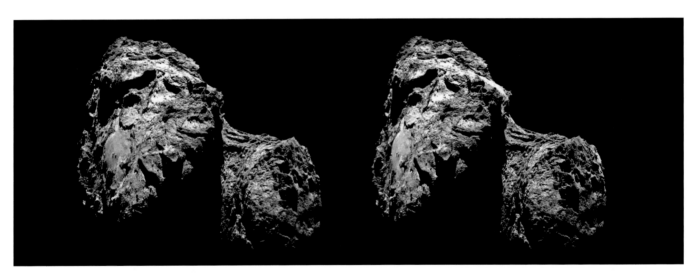

Comet 67P/Churyumov-Gerasimenko in depth – assembled from pictures taken by ESA Rosetta's OSIRIS narrow-angle camera
on November 22–23, 2015, when the Rosetta probe was about 130 kilometers (81 miles) from the comet's nucleus.

This stereo, the first ever 3-D close-up of Pluto, was assembled by Brian in the control room of the
New Horizons (NH) mission a few minutes after the second of these digitally transmitted views
arrived back on Earth. The NH probe was already speeding on towards the Kuiper Belt.

In 1985 the world witnessed Live Aid, with twin stages and sets in London, at Wembley Stadium, and in Philadelphia, at JFK Stadium. Again, the musical interlude served as a sort of gauge of where culture had gone from the Apollo era into the shuttle era. The acts made an incredible who's-who, including David Bowie, Paul McCartney, The Who, Led Zeppelin, Elton John, and Eric Clapton. But it was Queen who stole the show, in a 21-minute micro concert that ranged from "Bohemian Rhapsody" to "We are the Champions" — a mini concert in itself. In all fans of space, astronomy, and music, that day of Queen at Live Aid marks a special crossroads.

UNMANNED MISSIONS

Many missions of planetary exploration have blossomed over the past generation, since those heady days. Two recent missions have generated especially exciting results for fans of solar system bodies. Launched in 2004, the Rosetta spacecraft was a European Space Agency venture made to rendezvous with, explore, and even land a small probe on Comet 67P/Churyumov–Gerasimenko. In 2014 this amazing craft reached the comet, orbiting it and producing breathtaking pictures of the body, the best views of a cometary nucleus ever recorded. And the small Philae lander stunned the world by landing on the surface of the dirty iceball, providing incredible insight into an otherworldly celestial visitor.

Soon after the excitement over Comet 67P, the New Horizons spacecraft arrived at the Pluto system. Launched in 2006, the craft completed the original mantra of NASA's, begun in earnest with the Voyagers, to explore all the major bodies of the solar system. When the flyby occurred in 2015, the New Horizons team amazed the world with an incredible, vibrant, and far more active landscape on Pluto than anyone could have envisioned. Moreover, the close-up views of large moon Charon and peeks at the system's smaller moons completely rewrote our ideas about the outer solar system, the so-called Kuiper belt objects that exist out there in large numbers.

Other past missions that occurred between the Apollo program and the present day also added a great deal to our knowledge of the Moon. In 1994, the United States launched the Clementine spacecraft, a joint project of the Air Force and NASA, which tested spacecraft components during long exposure duration to spaceflight, to study the Moon, and to observe the near-Earth asteroid 1620 Geographos. Clementine's scientific objectives were aggressive, as it observed and imaged the Moon in visible light as well as infrared and ultraviolet wavelengths. The spacecraft operated for 115 days. It produced substantial laser altimetry on lunar features, as well as studying the Moon's gravitational field extensively. With these data, the spacecraft helped to create a multispectral imaging set of the entire Moon, as well as study lunar mineralogy.

In 1998, NASA launched its next substantial lunar project, the Lunar Prospector spacecraft. For 570 days this robust little craft operated in a low polar orbit of the Moon, collecting data on ice deposits around the Moon's poles, data on the Moon's gravitational and magnetic fields, and various outgassing events that occasionally take place on the Moon. To end Lunar Prospector's life, NASA crashed the probe into an area near Shoemaker Crater, close to the lunar south pole. Scientists hoped to detect the presence of water molecules released by the collision from Earth, but they were not able to make such a finding.

In 2009, NASA launched a substantial mission that is still in operation. The Lunar Reconnaissance Orbiter (LRO) is a spacecraft currently orbiting the Moon in an eccentric polar mapping orbit. It has been the powerhouse in terms of the current era of lunar mapping and imaging. Over the last decade the spacecraft has made a 3-D map of the Moon's surface at 100-meter (325-foot) resolution, covering some 98 percent of the lunar surface. The areas that have not been imaged are near the Moon's poles and lie in deep shadow. Among the numerous images collected by LRO are high-resolution pictures of the Apollo landing sites, to a resolution of 0.5 meters (1.6 feet). The mission was originally intended to last two years, and so it has outlived its projected life by a huge span.

Along with LRO, at the same time NASA launched the Lunar Crater Observation and Sensing Satellite (LCROSS), whose objective during its 144-day mission was to study water ice believed to exist in the permanently shadowed craters near the Moon's poles. This followed the detection of lunar water ice from the Chandrayaan-1 probe. LCROSS studied the lunar crater Cabeus near the Moon's south pole and detected water molecules in the crater after smashing a Centaur upper rocket stage into the crater, followed by an impact of LCROSS itself at mission's end.

The most recent American mission of significance was the Gravity Recovery and Interior Laboratory (GRAIL), which was built by the Massachusetts Institute of Technology and launched by NASA in 2011. Two small spacecraft, nicknamed Ebb and Flow, were launched on the same rocket and designed to impact the lunar surface. The objectives were to study the Moon's crust, understand its thermal evolution, place limits on the Moon's core size, and better understand the internal structure of the Moon. Following a long journey to the Moon, the two craft studied our neighbor in the spring of 2012, the mission lasting nearly three months. After collecting voluminous data, the craft were powered down and impacted the Moon near the craters Philolaus and Mouchez.

The Europeans also launched a lunar probe in SMART-1, a Swedish-designed European Space Agency spacecraft that blasted off for the Moon in 2003. The acronym represents "Small Missions for Advanced Research in Technology." This small probe carried a micro-imager, an X-ray spectrometer, an infrared spectrometer, and an electron and dust experiment. The craft detected calcium in Mare Crisium, returned close-up images of the lunar surface, and after nearly 3,000 orbits, crashed into the Moon in 2006, hoping to simulate a meteoroid impact and unleash water ice molecules. This event served as something of a "dry run" for the planned LCROSS impact.

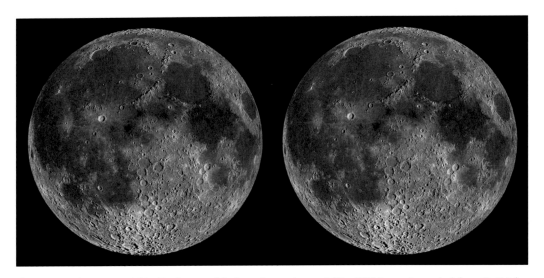

This 3-D Moon picture was published by the team of the Lunar Reconnaissance Orbiter (LRO) Camera in anaglyph format in October 2013, and we offer it here in parallel format. The images were generated by draping a mosaic of thousands of pictures taken by the LROC wide-angle camera when the Sun was low over a Global Lunar Digital Terrain Model. The low-Sun angle accentuates every topographical feature due to the shadows generated by the Sun being close to the horizon, bringing out the 3-D texture of the lunar surface.

THE GIANT IMPACT HYPOTHESIS

Many space enthusiasts lament the fact that no humans have walked on the Moon since the end of Apollo, nearly 50 years ago. Yet all of the lunar science that has occurred since then has revolutionized our understanding of the Moon. From the first days of analyzing the Apollo Moon rocks, scientists were astonished at their similarity to Earth rocks. In fact the specific gases that are trapped in Moon rocks are essentially identical to Earth rocks, in some very specific ways. The ratios of oxygen isotopes, the varieties of elements, that exist as gases trapped within the rocks, match those of Earth rocks. For a time this greatly puzzled scientists.

In the mid-seventies, planetary scientists William K. Hartmann and Donald R. Davis proposed an explanation for this. Their so-called "Giant Impact Hypothesis" has come to be widely accepted as the most likely idea about how the Moon formed. Planetary scientists believe that in the early history of the solar system, numerous bodies were impacting each other in the inner part of the system. All you have to do to appreciate this is to look at the surfaces of the Moon and Mercury to see a preserved record of ancient impacts. Large bodies, planetesimals, were also present in the ancient inner solar system. Hartmann and Davis proposed that a Mars-sized body, later called Theia by planetary scientists, impacted Earth some 4.53 billion years ago. This impact would have thrown out a ring of material that orbited Earth and accreted into the Moon. This would explain the similarity of Moon and Earth rocks and also a variety of physical attributes of the Moon itself. This idea is widely accepted today. Most of what was Theia was absorbed into Earth. Where did this ancient body go? You're standing on most of it.

So one of the boldest and most important ideas about our world and our closest celestial neighbor came from the Apollo program. The same evidence lies within lunar meteorites, pieces of the Moon that have been knocked into space by impacts and eventually fallen to Earth. I take pride in having a piece of one of these, Dar al Gani 400, in my house to remind me of the significance of the Moon to our world.

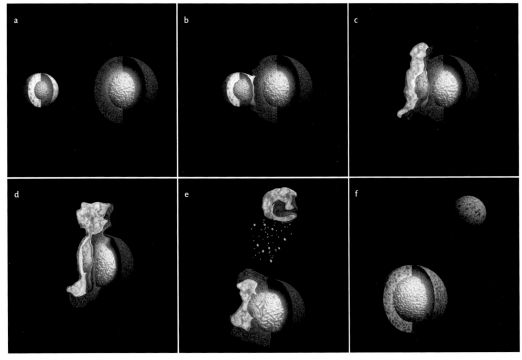

The Moon formed through the collision of two planets each with a mantle and a core is shown in the progression (a) – (f).
The cores coalesced into the larger planet, Earth, at (d) and (e), and the mantle debris became the Moon at (e) and (f).

THE STARMUS FESTIVAL

The friendships that ensued from the original astronaut and cosmonaut groups have much to say about human nature. The Apollo–Soyuz mission in 1975 was made famous by the handshake between Commanders Alexei Leonov and Thomas Stafford as the hatch between the Apollo and Soyuz capsules was opened. Their friendship has lasted a lifetime, and Alexei is godfather to Thomas's sons. Separately, Alexei Leonov wrote a fascinating book with Dave Scott, the Commander of Apollo 15, the seventh man to walk on the Moon, about their uniquely different experiences during the space race — *Two Sides of the Moon*.

In 2011, Garik Israelian, astrophysicist at the Institute of Astronomy, Tenerife, took this commonality of purpose and spirit between the astronauts and cosmonauts to the next level by hosting the Starmus Festival to celebrate 50 years since the first flight of Yuri Gagarin. A host of astronauts and cosmonauts, including Neil Armstrong and Buzz Aldrin, were able to enjoy each other's company during a week of talks and events in the Canary Islands, and the Festival has gone from strength to strength in subsequent years. Here are some of the photographs from the Festival.

Neil Armstong in deep conversation with Alexei Leonov at Starmus 2011.

Bill Anders at Starmus 2011.

Walt Cunningham signing books at Starmus 2014, with (front to back):
Charlie Duke, Garik Israelian, Harold Kroto, Brian May, and David Eicher.

Buzz Aldrin delivers his account of the Apollo 11 mission at the very first Starmus, 2011.

Neil Armstrong at Starmus 2011.

Alexei Leonov using a blackboard to tell the story of his first spacewalk.

Harrison Schmitt.

Charlie Duke.

Alexei Leonov sketches for Garik Israelian's son Arthur at Starmus 2011.

AFTERWORD
BY JIM LOVELL

My adventures with the Moon began in 1958, when, as a Navy pilot, I was selected as one of the astronaut candidates for the Mercury program. A medical glitch kept me out of that group, but in 1962 I made it into the "New Nine," as NASA called astronaut group 2. I soon served as backup pilot for the Gemini IV mission, and in December 1965 I flew in my first spaceflight, along with Frank Borman, in Gemini VII. A second flight, Gemini XII, paired me with Buzz Aldrin in November 1966, and we made 59 orbits, docked with an unmanned vehicle, and Buzz made three spacewalks as I piloted the craft.

The race to space was a thrilling time, and it was an incredible experience to be part of this era. I was delighted to see that Brian May and his friends have assembled this new and unique book about the Apollo era and the space race with such a wide range of stereoscopic pictures. Nothing has been produced like this before, and I believe it will show the Apollo era in a new and interesting way. I have been a guest and participant at the Starmus Festival, and happy to spend time with Garik Israelian, and the whole Starmus gang, including fellow space travelers Alexei Leonov, Charlie Duke, and others, and old friends like Neil Armstrong and Buzz Aldrin.

The images carry me back strongly to the days of Apollo. I was privileged to serve as Command Module Pilot for Apollo 8, the first big "test run" that carried our spacecraft around the Moon and back, practicing and perfecting the techniques that would land Neil and Buzz a few months later. Frank Borman, Bill Anders, and I circled the Moon, gazing down on it from a distance of a little more than 100 kilometers (60 miles). I described the view of the Moon as "essentially gray, no color; looks like plaster of Paris or sort of a grayish beach sand. We can see quite a bit of detail. The Sea of Fertility doesn't stand out as well here as it does back on Earth."

It was an amazing experience to fly around the Moon. And then I had the fortune to be chosen as part of the crew of Apollo 13. Space enthusiasts know the story of that mission; en route to the Moon, one of our oxygen tanks in the Service Module experienced an explosion, and the situation threatened our lives. The world watched. We aborted the lunar landing. We had to fly around the Moon and circle back to Earth, using the Lunar Module as a lifeboat. The experience has been dramatized in a famous movie and in books. Living the experience was not easy, but it placed the value of life, and the fragile nature of our existence on Earth, in deep perspective.

Jack Swigert, Fred Haise, and I survived our brush with the cosmos and returned to Earth safely. Americans went on to send four more missions to explore our nearest celestial neighbor. And we have not been back since.

I hope that this book – with David J. Eicher's brilliantly evocative text and Brian May's sensational stereo pictures – will give you something of the experience of those days of the space race to the Moon. May we as explorers live to see another great age of discovery of this magnificent universe we live in.

November 12, 1966: astronaut Jim Lovell is photographed inside his Gemini spacecraft during the Gemini XII mission.

June, 2011: Jim Lovell photographed at Starmus on Tenerife.

GLOSSARY

Agena Target Vehicle
An unmanned US space vehicle used for docking practice.

Albedo
A measure of how reflective a surface is.

ALFMED
Apollo Light Flash Moving Emulsion Detector, an experiment designed to identify flashes of light witnessed by closed-eye astronauts, which came from cosmic ray strikes.

ALSCC
Apollo Lunar Surface Close-up Camera, a stereoscopic camera carried by the astronauts to photograph small areas of the lunar surface.

ALSEP
Apollo Lunar Surface Experiments Package, a suite of scientific experiments made by Apollo 12 and later missions.

Anaglyph images
A method of making red and green images to simulate stereo pictures.

Anorthosite
Rock consisting of *plagioclase feldspar*.

Apollo 1 tragedy
A flash fire that killed the crew of Apollo 1 during a 1967 test.

Apollo program
Also called Project Apollo, the third US manned spaceflight program, lasting from 1966–72.

Apollo-Soyuz Test Project
A cooperative joint flight between the Americans and the Soviets in 1975.

Baikonur Cosmodrome
The principal Soviet and Russian launch facility, located in southern Kazakhstan.

Basalt
Volcanic rock.

Bay of Pigs
An ill-fated US invasion of Cuba in 1961.

BIOCORE
A biological experiment on Apollo 17 containing mice.

Breccia
Rock consisting of fragments, sometimes recrystallized together.

CAPCOM
CAPsule COMmunicator: a communications specialist at the Mission Control Center designated to talk with the astronauts.

Capsule communicator
See CAPCOM.

Circumlunar
Around the Moon.

Clementine
A US robotic lunar mission that flew in 1994.

Command Module
The Apollo orbiting craft that circled the Moon as the lander descended.

Cosmic rays
High-energy particles that fly through space.

CSA
Canadian Space Agency.

CSM
Command/Service Module, the Apollo spacecraft component that orbited the Moon as the lunar lander descended to the surface.

Cuban Missile Crisis
A period in late 1962 marked by a showdown over Soviet missiles deployed in Cuba.

EASEP
Early Apollo Scientific Experiment Package: a suite of science experiments carried on Apollo 11.

"Eight Ball" Attitude Indicator
Instrument to show the orientation of the spacecraft, named for its resemblance to a pool ball.

Electrophoresis
A study of the motions of molecules in a fluid under an electrical field.

ESA
European Space Agency.

Extravehicular activity
A venture by an astronaut in space or on the Moon, outside of the craft, i.e. a spacewalk or a moonwalk.

Fallen Astronaut Memorial
An aluminum sculpture made by Paul Van Hoeydonck left on the Moon during Apollo 15 to commemorate deceased astronauts and cosmonauts.

Gemini program
Also called Project Gemini, the US manned spaceflight program lasting from 1964–6.

Genesis Rock
A rock, thought to be very primitive, recovered during Apollo 15.

Giant Impact Hypothesis
The most widely accepted hypothesis for the formation of the Moon, suggesting the long-ago collision of a Mars-sized body with Earth, throwing out a ring of material that accreted (coalesced) into the Moon.

GRAIL
Gravity Recovery and Interior Laboratory, a NASA lunar science mission 2011–12.

International Space Station (ISS)
A multi-country orbiting space station launched in 1998 and continuing to operate.

J-type mission
Apollo missions designed to spend longer than three days on the lunar surface.

JAXA
Japan Aerospace Exploration Agency.

Kennedy Space Center
A spacecraft launch facility built at Cape Canaveral, Florida.

LCROSS
Lunar Crater Observation and Sensing Satellite, a 2009 NASA mission to study the Moon.

LM
Lunar Module, the craft that carried Apollo astronauts to the lunar surface and back.

Luna program
A Soviet unmanned lunar exploration program lasting from 1959–76.

Lunar Module
The Apollo component that descended to the Moon's surface.

Lunar Prospector
A NASA mission that explored the Moon 1998–9.

Lunar Reconnaissance Orbiter
A NASA spacecraft launched to study the Moon in 2009 that continues to the present.

Lunar Roving Vehicle
The "Moon buggy" cars taken along on Apollos 15, 16, and 17 that helped astronauts explore wider lunar areas.

Lunokhod
A Soviet program that landed two unmanned rovers on the Moon's surface.

Magnetometer
An instrument used for measuring magnetic forces.

Mariner program
A US robotic exploration program of Mars, Venus, and Mercury, from 1962–73.

Marshall Space Flight Center
A rocketry engineering facility in Huntsville, Alabama.

Mass spectrometer
An instrument that measures the masses within a sample.

Mercury program
Also called Project Mercury, the first US manned spaceflight program, lasting from 1959–63.

Meteoroid
A small piece of rocky debris that can strike solar system bodies or spacecraft.

Microbreccia
Breccia containing relatively small particles of mineral or rock.

Mir
A Soviet/Russian space station that operated in low-Earth orbit from 1986–2001.

Mission Control Center
A spacecraft control complex built in Houston, Texas, and later renamed Johnson Space Center.

Modular Equipment Transporter (MET)
A wheeled cart used on Apollo 14 to move equipment along the Moon's surface.

NASA
National Aeronautics and Space Administration, the US space agency.

Nedelin catastrophe
A 1960 launch pad accident at the Baikonur Cosmodrome that killed at least 78 people.

New Horizons
A NASA spacecraft launched in 2006 that conducted a flyby of Pluto and its moons in 2015.

OWL viewer
A stereoscopic device invented by Brian May (and included in this book!) that allows viewing of stereo images.

Plagioclase
A group of *tectosilicate* minerals belonging to the feldspar group.

Pogo oscillations
Vibrations in a spacecraft that can cause various problems.

Porphyritic basalt
Volcanic rock with a large range of crystal sizes.

Ranger program
A US robotic lunar exploration program that lasted from 1961–65.

Regolith
Loose, rocky material on the lunar surface.

Rille
A fissure or narrow channel on the Moon's surface.

Roscosmos
Roscosmos State Corporation for Space Activities, the Russian Space Agency.

Rosetta
An ESA spacecraft launched in 2004 that visited the comet 67P/Churyumov-Gerisamenko from 2014–16 and deployed the Philae lander onto the comet's nucleus.

S-band
A part of the electromagnetic spectrum from 2-4 gigaherz (GHz).

Salyut program
A Soviet program, commencing in 1971, that launched the first-ever space station.

Saturn V
The primary booster rocket of the Apollo program.

Scanning Electron Microscope (SEM)
A high-powered microscope that uses a beam of electrons to form a reflected image.

Sequential stereoscopy
The creation of two images separated by time or distance to make a 3-D picture.

Service Module
A portion of the Apollo spacecraft containing support systems.

SMART-1
An ESA lunar probe that operated from 2003–6.

Soyuz
The third Soviet manned space exploration program, lasting from 1967 onward.

Soyuz 1 tragedy
A 1967 accident that killed Vladimir Komarov on reentry.

Soyuz 11 tragedy
A 1971 Soviet mission that tragically resulted in the deaths of the three-man crew.

Space Shuttle
A US manned exploration program in low-Earth orbit that lasted from 1981–2011.

Spectrograph/Spectrometer
An instrument used to view and analyze wavelengths in the electromagnetic spectrum.

Star City
The nickname given to the Soviet cosmonaut training facility in Moscow Oblast, Russia.

Starmus Festival
A science festival founded by Garik Israelian in 2011.

Stereopsis
The ability to see images in three dimensions.

Stereoscope
A device made to show two stereoscopic images as a 3-D picture.

Stereoscopy
The creation of two images that can blend together to make a 3-D scene.

Surface Electrical Properties Experiment
A suite of devices carried along on Apollo 17.

Surveyor program
A US unmanned lunar exploration program that lasted from 1966–68.

Tectosilicate
Silicate minerals containing crystals with a three-dimensional framework of tetrahedra.

Terminator
The dividing line between the illuminated and unilluminated parts of a planet or moon.

Trans-lunar injection
A spacecraft propulsive maneuver that sets it on a course to orbit the Moon.

V-1 rocket
A German rocket employed during World War II.

V-2 rocket
A sophisticated German rocket employed during World War II.

Van Allen belts
Regions of intense radiation surrounding Earth.

Voskhod program
The second Soviet manned spaceflight program, lasting from 1964–65.

Vostok program
The first Soviet manned spaceflight program, lasting from 1959–63.

Zodiacal Light
A glow caused by sunlight scattering off tiny dust particles in the plane of the solar system.

Zond program
A Soviet unmanned lunar exploration program that lasted from 1964–70.

CLAUDIA MANZONI'S QUICK STEREOSCOPIC GUIDE TO ALL SIX APOLLO LANDING SITES!

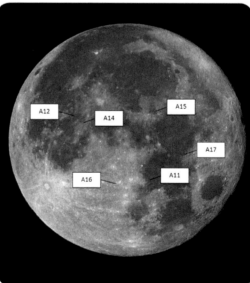

A11 (Apollo 11): Mare Tranquillitatis
A12 (Apollo 12): Oceanus Procellarum
A14 (Apollo 14): Fra Mauro

A15 (Apollo 15): Hadley/Apennines
A16 (Apollo 16): Descartes
A17 (Apollo 17): Taurus-Littrow valley, Mare Serenitatis

Based on full Moon photographs taken by astronauts on board the ISS (right: May 9, 2009 - left: July 7, 2009).

ACKNOWLEDGEMENTS

We would like to acknowledge the following for their generous help and support:

Katie Helke (MIT), Gill Norman, Suszann Parry, Bert Ulrich, Andrew Chaikin, Rich Dubnow, Cindy Slocki, Ken Kong, Tom Jackson, James Pople, David Burder, Matt Taylor (Rosetta), Alan Stern (New Horizons), Joel Parker (LRO), Phil Webb, Pete Malandrone, Nigel Burchett, Noah Petro, Katherina Gauss, Sharon Ashley, Paul Schenk, Dave Dunlop, and Wolfy.

In particular, we would all like to acknowledge the eternal inspiration of the late Sir Patrick Moore, who spent a lifetime studying the Moon, and was consulted by NASA on lunar topology and the potential hazards presented by lunar dust. Through us, we see this book as being part of his legacy, in the continuing spirit of curiosity and limitless possibilities.

CREDITS:

The London Stereoscopic Company claims copyright in all the stereoscopic images in this book. The majority were derived from NASA orginals and the authors' own collections, unless otherwise specifically stated.

However, despite extensive searches, we have been unable to determine with 100% certainty the origins of some of the pictures. The publishers would welcome any input to enable us to credit the appropriate sources.

Abbreviations: Brian May (BM), Claudia Manzoni (CM), Denis Pellerin (DP)

Mono credits

NASA p. 2, 4, 5, 12, 25, 28, 29, 30 (top), 37, 38 (top), 40,   46 (bottom), 49, 50 (top and bottom), 52, 55 (bottom), 56, 57 (bottom), 58, 60, 63, 70, 72, 75, 79, 82 (top), 84 (top and bottom), 85, 90 (top and gutter), 94 (bottom), 97 (bottom), 98 (bottom), 100 (top and bottom), 101 (top and bottom), 102, 103, 105, 106, 107, 109, 114, 116, 120, 124, 126 (top and bottom), 127 (bottom), 128, 130 (top), 132 (bottom), 133, 134 (top), 136 (bottom), 137 (top), 140, 142, 143, 146, 148, 151, 152, 154, 155 (top and bottom), 159, 160, 162 (top and bottom) , 165 (bottom), 168, 172, 173 (top), 187 (background)
Max Alexander p. 15 (top)
DP p. 15 (bottom)

Science Photo Library/Sputnik p. 16 , 22 (top and bottom), 41 (bottom)
David J.Shayler/Astro Info Service Ltd collection p. 23, 35, 42, 43 (right), 137 (bottom)
Internet Archive (Archive.org) p. 24 (top)
CIA/Public Domain p. 26, 27
Deutsches Bundesarkiv p. 31
Alexei Leonov p. 43 (left)
Robin Rees p. 51, 122
Public Domain (Wikipedia) p. 53, 91 (top)
Jodrell Bank Observatory p. 54, 55
Getty Images p. 64
Alan Bean p. 104
Roscosmos p. 138, 139
James Symonds p. 181

Stereopicture credits

NASA/JSC – William3D – BM: p. 7 (top), 112 (top), 158

DP – BM: p. 7 (bottom), 9, 10 (top), 14, 184

BM: p. 10 (bottom)

View-Master: p. 11, 32 (bottom), 59

Science Museum – CM – BM: p. 19

Roscosmos – BM – CM: p. 18, 23, 24 (bottom), 44

NASA/JSC – CM – BM: p. 12, 62, 74, 76, 77, 78, 82, 83, 85, 86, 87, 88, 89, 90, 92, 93 (top and middle), 95 (bottom), 96 (bottom), 97, 99 (top), 100, 110, 111, 112 (bottom), 113, 115, 118 (bottom), 119, 121,122, 125, 131, 132, 134, 136, 145, 146, 147, 148, 149 (top), 150 (top), 157, 161, 163, 167, 169, 170, 171

CM – BM: p. 13, 17, 32 (top), 41, 54, 61, 69, 139

Chiara Tomaini – BM: p. 20

Archive.org video – CM – BM: p. 24 (top)

NASA/MSFC – BM: p. 30 (bottom)

NASA video – CM – BM: p. 33 (top), 46, 118 (top)

British Pathé video – CM – Brian May: p. 33 (bottom), 57

NASA – William3D – BM: p. 38, 127, 150 (bottom), back cover

Steven Young – BM: p. 45 (middle)

NASA – CM – BM: p. 45 (bottom), 173, 174, 175, 176 (middle)

NASA/JSC/Arizona State University – CM – BM: p. 47, 48

NASA/USGS/LPI – CM – BM: p. 63

Hancock County Chamber of Commerce – CM – BM: p. 73

Smithsonian Institution – CM: p. 93 (bottom), 94

NASA/GSFC/Arizona State University – CM – BM: p. 95 (top), 108, 130, 180

NASA/JSC – David Burder – BM: p. 96 (top)

NASA/JSC – DP – BM: p. 98, 99 (middle and bottom), 156, 166

Lehigh University, Bethlehem, Pennsylvania: p. 149 (bottom)

ESA/NASA – CM – BM: p. 176 (bottom)

Canadian Space Agency – CM – BM: p. 177

ESA/Rosetta/MPS for OSIRIS Team MPS/UPD/LAM/IAA/SSO/INTA/UPM/DASP/IDA – CM – BM: p. 178 (middle)

NASA/Johns Hopkins University Applied Physics Laboratory/Southwest Research Institute – BM: p. 178 (bottom)

BM: p. 178 (top), 182, 183 (middle, bottom), 185, 187

Terry Briscoe — BM: p. 183 (top)

NASA/JSC Earth Science and Remote Sensing Unit – CM: p. 189

Stereoscopic Viewing

If you're familiar with the technique of **parallel free-viewing** of 'side-by-side' stereo pairs, you will have no trouble seeing all the 3-D images in this book. But to get the best immersive effect, use the OWL viewer supplied here.

This viewer is the new hand-held version of the acclaimed London Stereoscopic Company OWL Stereoscope.
It's designed to give you high quality 3-D viewing of the stereo pairs in these pages.
With careful use of this deceptively simple device, you will experience
the journey of the Moon Landings through over 150 new stereo views
in a depth and realism never before seen.

Directions

Remove your LITE OWL viewer from the pocket opposite, and hold it with
either hand, locating the thumb on one of the grips near the lower corners.
Position the book in good light, with no shadows on the page.
Bring the lenses of the viewer close to your eyes, and position your
head squarely in front of the image pair you've chosen to view.
Move to a distance of about 5 inches (120 mm) from the page, and relax the eyes.
Don't squint or strain – you won't need to – and forget the idea
that you're viewing something close-up. Instead, expect to witness
a view of a distant scene through a window, or through binoculars.

Gently move back and forth a few millimeters until the image is in focus,
and slightly adjust the tilt of your head if necessary to align it with the page.
At this point the eyes should settle into a relaxed position as the full beauty of the
three-dimensional image appears. If you still see two flat images, don't give up!
Just disengage for a few seconds and look around the room, to remind yourself what
that feels like, and then try again, remembering to relax and look into the distance.
It will be worth it!

For more help, and information on all kinds of stereoscopic matters ancient
and modern, please visit our London Stereoscopic Company website at

www.LondonStereo.com

THE
London Stereoscopic Company,
LTD.